The SECRET LIVES of DINOSAURS

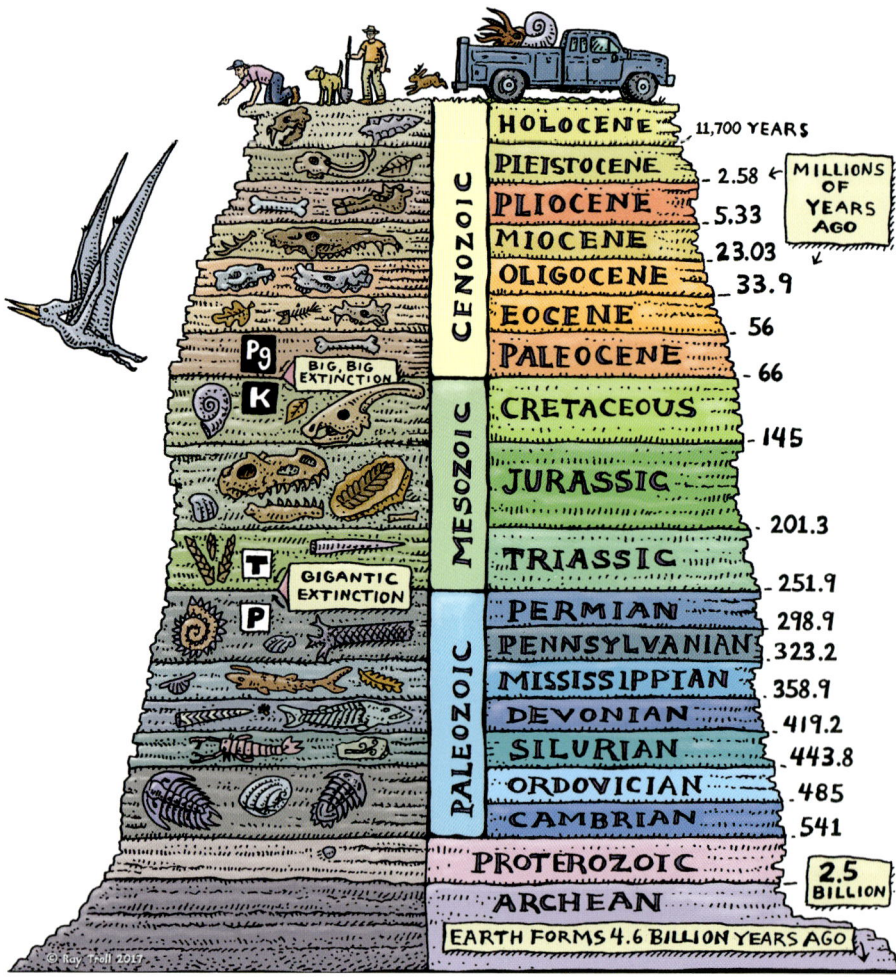

Era	Period/Epoch	Millions of Years Ago
CENOZOIC	HOLOCENE	11,700 YEARS
	PLEISTOCENE	2.58
	PLIOCENE	5.33
	MIOCENE	23.03
	OLIGOCENE	33.9
	EOCENE	56
	PALEOCENE	66
MESOZOIC	CRETACEOUS	145
	JURASSIC	201.3
	TRIASSIC	251.9
PALEOZOIC	PERMIAN	298.9
	PENNSYLVANIAN	323.2
	MISSISSIPPIAN	358.9
	DEVONIAN	419.2
	SILURIAN	443.8
	ORDOVICIAN	485
	CAMBRIAN	541
	PROTEROZOIC	2.5 BILLION
	ARCHEAN	

MILLIONS OF YEARS AGO

BIG, BIG EXTINCTION

Pg
K

GIGANTIC EXTINCTION

T
P

EARTH FORMS 4.6 BILLION YEARS AGO

© Ray Troll 2017

The
SECRET LIVES
of DINOSAURS

Unearthing the *Real* Behaviors
of Prehistoric Animals

DEAN R. LOMAX

Illustrated by Bob Nicholls

Columbia University Press
New York

Columbia University Press
Publishers Since 1893
New York   Chichester, West Sussex
cup.columbia.edu

Library of Congress Cataloging-in-Publication Data
Names: Lomax, Dean R. author | Nicholls, Bob, 1975– illustrator
Title: The secret lives of dinosaurs: Unearthing the real behavior of prehistoric animals /
   Dean R Lomax ; illustrations by Bob Nicholls.
Description: New York : Columbia University Press, 2025. | Includes bibliographical
   references and index.
Identifiers: LCCN 2025004030| ISBN 9780231211307 hardback |
   ISBN 9780231558846 ebook
Subjects: LCSH: Animals, Fossil | Animals, Fossil—Behavior | Animal behavior—
   Evolution
Classification: LCC QE761 .L663 2025 | DDC 560 23/eng/20250—dc18
LC record available at lccn.loc.gov_2025004030

Printed in the United States of America

Cover image: Bob Nicholls
Frontispiece: Ray Troll

GPSR Authorized Representative: Easy Access System Europe, Mustamäe tee 50,
10621 Tallinn, Estonia, gpsr.requests@easproject.com

To Judy Massare and Bill Wahl, thank you for recognizing the potential of an eighteen-year-old Yorkshire lad with a fascination for fossils. Your encouragement and guidance have helped me to become the paleontologist I am today.

For that, I will be eternally grateful.

And to Levi Shinkle, thanks for all your support. Also, for putting up with all those years of marine reptiles. Blame Judy and Bill!

. . . Oh, and to my family for their ever-present support and for trying to show me that there is more to life than playing with dinosaurs and other prehistoric animals. There isn't.

Contents

Prologue: Paleo in Perspective

It's 3:07 a.m. My eyes quickly shoot open, triggered by the sound of scratching around my head. I lie silent, wrapped inside my sleeping bag, cramped inside a tent too small for my long body. To combat the freezing conditions, I am wearing three layers of clothing, including a woolly hat. I listen carefully. Nothing. All I hear is the tranquil, calming noise of rain gently striking the tent and the occasional whisper of the wind. Suddenly, something brushes past my foot against the edge of the tent. I comically jump up and hit my head on the low roof as I hastily hunt for my head torch. *Flick*. I spot the culprits—the distinctive silhouette of mice scurrying around the outside of the tent, attempting to work their way in, craving the warmth. I rather unconvincingly tell myself that the tent will hold up. Right?

Morning rolls around, and I wake to the telltale traces of mice—bite marks galore, mixed with a smattering of poop and pee, sitting on the mesh netting *inside* the tent. Little tidbits of evidence that these mice burst through the first barrier but could not quite make it through the second. Phew.

In April 2022, this was my life for two weeks camping in the wilderness of Torres del Paine National Park, in southern Chilean Patagonia. Or, as several of the locals called it, "the arse end of nowhere." I was invited to join a small team of paleontologists led by my friend and colleague Judith Pardo-Pérez. Our mission was to recover the remains of "Fiona" from near the side of a massive glacier. No, not a human left behind on some previous expedition, but a pregnant, ancient marine reptile, an ichthyosaur. A five-hour horse ride or a ten-hour hike is required to get to the dig site. It is so remote that fewer than fifty people on the planet have ever stepped foot there. Not only was I there to excavate this important find but also to record an exciting documentary that, with a little luck, you might get to watch one day.

One freezing evening, as I lay watching the devoted mice plot their next attempt to make it inside, my mind began to wander to the sound of the rain pummeling down, a noise that has been ever present on our planet. It is fascinating to think about how something as ephemeral as the sensation of rain—its sound, smell, and feel—can connect us across time and space to other organisms from a bygone era that no doubt experienced similar sensations. There I was, having traveled halfway around the world to one of the most remote and extreme fossil locations on the planet. Dedicating weeks of my life while perched beside an immense glacier, without proper food, a toilet, or a shower, with unsettling reports of a one-eyed cougar, and entirely cut off from the world. All to dig up an ancient marine reptile in the name of science.

My little story is just one of the countless tales that paleontologists and fossil hunters experience on their adventures. It is amazing what we are willing to do for the love of paleontology. Sometimes, it is easy to forget that so much sacrifice, time, and dedication goes into rescuing and studying these incredible creatures from the depths of time. Is it worth it, you might ask? Is it worth the often extreme, physically and mentally challenging conditions, the rations of food, or the occasional mouse attack, all for a rocky old fossil? You bet it is.

All of this got me reminiscing about 2008 and a summer Sunday fossil hunt at a site in Wyoming. I had been in the United States for a few months at this point, volunteering at the Wyoming Dinosaur Center, a museum located in the heart of dinosaur country. One of the students invited me and some other volunteers to look for marine fossils at a nearby site. During our little excursion, we were joined by Bill Wahl, a paleontologist and marine reptile expert. He had been telling me all week about this fossil site and explained that we could find all sorts of ancient marine creatures.

After a few hours of hunting in the dry, baking heat, we took a break next to some ant mounds. Bill, as sharp as ever, shouted me over and asked me to look closely at the ant mounds. Naively, I figured he was meaning I should look for ants. Not quite. I got in close and stared carefully at the mounds, and what did I see? Fossil-collecting ants. The mounds contained numerous tiny fossils that would be easily missed by the naked eye. There were swaths of shark teeth, ancient cephalopods, and fish scales, among numerous other fossils. I quipped that these ants were better fossil hunters than me!

That summer in Wyoming marked a significant turning point in my career. I spent a lot of time chatting with Bill about fossils, and, serendipitously, he introduced me to another paleontologist and marine reptile expert, Judy Massare, a New York professor who was visiting the area. If you have followed my academic research or even just my adventures on social media, then it is highly likely that you will know my area of expertise is ichthyosaurs, which is all thanks to this chance meeting. These wondrous denizens of the prehistoric seas are what some (like me) might say are the coolest of all the so-called "sea dragons"—and the same such creature that we were unearthing in icy Patagonia.

I have spent most of my professional career—at the time of writing this—of seventeen years studying thousands of ichthyosaurs across the globe, among many other types of fossils. I grew up learning about these ancient reptiles from books and visiting museums. Britain is the birthplace of ichthyosaurs. Their fossils have been found here for more than two hundred years, going back to the earliest scientific discoveries by the marvelous Mary Anning on the Jurassic Coast. Yet, rather surprisingly, my academic journey into ich-thyosaurs began in my hometown of Doncaster in Yorkshire, England.

Following the successful trip to dig for dinosaurs (and take tips from fossil-finding ants) in Wyoming, I began volunteering at Doncaster Museum in September 2008. I was very grateful to the curator, Peter Robinson, who saw my potential and offered me this experience. On my first official day, I was introduced to what was supposedly a "really good ichthyosaur replica." However, I immediately realized that this was a genuine fossil and was prob-ably scientifically important. The skeleton, originally from Dorset's Jurassic Coast in England, was just over a meter long and exposed a beautiful skull equipped with toothy jaws, but my eyes were drawn to a dark mass between the ribs: the ichthyosaur's last meal.

How fortuitous. Just a few months before, I met with two renowned ichthyosaur experts. Taking me under their wing, Judy and Bill encouraged me to write up the ichthyosaur for an academic journal. I will confess that I was clueless about how to do that, but I eagerly began the research.

After countless drafts of what eventually started to resemble an academic paper, following support from Judy, Bill, and others, I described this ichthyosaur with its last meal preserved. That dark mass was packed full of tiny, hook-shaped objects (called hooklets) from the arms of squid, along with a single fish scale. Some 187 million years ago, in the Early Jurassic, this ichthyosaur was feasting on squid-like cephalopods and fish before it died. Forming my first-ever scientific publication, the research was published in September 2010 in a journal in New York. That was it. I was officially a published scientist.

It would be remiss of me not to mention that I went on to study this ichthyosaur in more detail with Judy. Together, we determined that it was a new species, which we named after Mary Anning—*Ichthyosaurus anningae*, the one and only ichthyosaur named after the "mother of paleontology." I should probably also say that a little later in my career, I described a new genus and species of ichthyosaur that I named in honor of my two amazing mentors: *Wahlisaurus massarae*.

Unknowingly, at the time of my first study, I was setting myself up for a lifelong fascination with fossils that record direct evidence of behavior. Having spent that summer in Wyoming, meeting world-leading scientists and seeing other unexpected fossils with evidence of behavior, the idea was etched in my mind, just like those fossil-finding ants. Those ants did not know they were collecting fossils; they simply went about their day, building their homes. Similarly, those pesky Chilean mice were just looking for warmth, shelter, and food, and they left their traces in the process.

Each decision that you or any animal has ever made in the history of life is evidence of action. Whether you are trying to find food, make your home, or travel far and wide to follow your dreams, every decision in life is an act of behavior and a choice you make. Some good, some bad.

In this book, we dive deep into the fragility of the fossil record in preserving a repertoire of fossils with unusual or even extreme acts of behavior, challenging what we thought we knew about day-to-day life and survival in the prehistoric world.

The SECRET LIVES of DINOSAURS

Introduction

The Journey Doesn't End Here

Thanks for picking up this book. I hope you will not regret your act of *behavior*. Chances are, if you are reading this, then you might be here because you read one of my previous books, *Locked in Time*, which will act as the perfect primer for this one. You might even consider this a sort of sequel. But, hey, if you have yet to read *LIT*, then fear not—this book will take you on a new adventure. When I wrote *Locked in Time*, I had been developing the idea for a book dedicated to prehistoric animal behavior ever since that fateful trip to Wyoming in 2008. I had always hoped the idea could evolve into much more, and I am thrilled that you are joining me on this journey.

As a youngster growing up with an extreme obsession with fossils, I would do everything in my power to learn about the ancient world. My family always had pets, including dogs, rabbits, and fish, along with the occasional snail or ant that I would collect from the garden. For whatever strange reason, being surrounded by animals made me feel closer to the long-dead dinosaurs and other animals that fueled my love of paleontology.

The single best way to learn about an animal is by observing and studying its behavior. Simply go for a walk, and you will no doubt see some birds fluttering above, perhaps squabbling over food, or you might spot a curious squirrel zipping up the tree and caching its nuts. Maybe you notice the intricacy of a spider's web shimmering in the sunshine, with a few flies entangled. A fly caught in a spider's web is one of those interactions with which we are all familiar, something so seemingly simple yet inherently complex. The careful craft of the intricate web, the stealthy patience of the spider, and the unlucky fly that did not pay attention. Have you ever noticed any of these things? If you have, then great job. You have taken your first steps into *ethology*, the science of animal behavior.

Researching and understanding behavior in living animals provides paleontologists with key observational skills to help interpret behavior in fossils. This is *paleoethology*, which can provide vital insight into and have far-reaching implications for our understanding of the interrelationships between ancient animals and their environments, otherwise known as *paleoecology*. Crucially, for a paleontologist, or more precisely a paleoethologist, you become the ultimate detective and must switch between the world of the living and the very long-dead past.

The nature of paleontology is built upon the direct result of chance, often of an ancient animal being in the wrong place at the *right* time. Paleontologists must approach the study of a fossil with great care and patience, a bit like a CSI-style investigation—call it fossil forensics if you like. With fossils, we know how it ends. Everything died, so skip to the end. But what happened during the animal's life is where the *real* action is. Or, as in the wise words of Gandalf from *The Lord of the Rings*, "The journey doesn't end here. Death is just another path, one that we all must take." The unfortunate, albeit timely death of a creature from the past laid the path of discovery literally at our feet.

There is widespread public fascination with fossils that record the real-life stories of ancient animal behavior. A pair of dueling dinosaurs, a curious creature found at the end of its final footsteps, or a mother caring for her babies—these discoveries hit the headlines because they are exceedingly rare and because behavior breathes life into the vast prehistoric past. We also love the drama. In many ways, it makes these animals much more relatable and real. Despite the often multimillion-year differences separating you and that rocky fossil, these familiar actions show us that time has no barrier.

In the following eleven chapters, prepare to buckle up as we hurtle hundreds of millions of years into the past, charting the life cycle of prehistoric animals as told through some of the most extraordinary fossils. Capturing their journey, these chapters cover the story of life, from conception to demise, culminating with a chapter documenting some of the most unusual fossils ever. Establishing the theme and illustrating the intricacy involved in interpreting behaviors in fossils, each chapter is briefly introduced based on a bizarre behavior observed in the modern world.

To set the tone, did you ever hear about the weasel that was observed riding on the back of a flying woodpecker? As unlikely as this might sound, a photographer snapped this dramatic moment and witnessed the brief phenomenon unfold. Perhaps it was an attack gone horribly wrong on the weasel's part, though both survived the ordeal. Capturing this instant in real time was exceptionally rare, so imagine trying to explain such an occurrence in the fossil record.

Naturally, we *know* that we will never know all the ins and outs of the behavior of ancient animals because we were not there to observe them in real time; there are no Sir David Attenborough documentaries for us to watch in wonder. Yet, this is one of the reasons why we are so captivated by prehistoric life. But rather than speculation or pure fantasy, remember that everything in this book is based on evidence and comprehensive research. That includes the incredible color illustrations by Bob Nicholls, who beautifully brings these stories to life. Along the way, I share various anecdotal tales and other observations from my own adventures, helping to guide you on this deep-time journey.

Dinosaurs will forever be the shining emblem of the paleontology world. It is how it is. I know this, you know this, and the whole world knows this. Expectedly so, dinosaurs feature heavily in this book—no surprise there— but they are joined by a myriad of other exquisite animals that came long before and after the dinosaur's reign.

These fossils offer an unparalleled insight into the private lives of ancient animals. They allow us to appreciate and understand how even the most downright bizarre and often complicated forms of behavior can find their way into the fossil record given the right conditions. The astonishing evidence will take you on a roller coaster of emotions. Some fossils will surprise you, while others might make you laugh or even shed a tear. You might even develop a sense of empathy for these long-dead creatures, almost as if they were your own pets, helping you connect with the ancient world in a way and on a level that you may have never thought possible. Enjoy the ride as we journey into the lost worlds of prehistoric animals, unearthing life's eternal secrets forever written in the rocks.

1 | Starting Out

Several species of whiptail lizards are entirely female and have an unusual reproductive strategy. To stimulate reproduction, some of them engage in pseudocopulation, taking turns mounting each other to promote fertilization. This leads to reproduction by parthenogenesis, meaning that their eggs develop into embryos without fertilization from a male. All the resulting eggs will be female.

THE ROOTS OF REPRODUCTION

Did it all start with a bang? Well, theoretically, the universe began with a scorching hot *Big Bang* around 13.8 billion years ago, so that is one way of looking at it. However, life began in the ocean, and so did reproduction. The oldest unequivocal evidence of life comes from the Precambrian, around 3.5 billion years ago, an almost unimaginable expanse of time. Small in size and basic in structure, simple, single-celled bacteria paved the way for the evolution and diversification of life.

Much of what we know about life during this time comes from the remains of stromatolites or "layered rocks." Having the appearance of rocky mounds or columns, stromatolites are formed by colonial microorganisms, particularly slimy photosynthesizing bacteria (often called "blue-green algae," these cyanobacteria are not actually algae at all), sandwiched between layers of mud and built up over time. They are especially important in the grand scheme of life because they helped to pump oxygen into the early atmosphere, which paved the way for the evolution of complex life forms.

These mounds of slime may once have ruled the primordial world's shallow oceans, but stromatolites can only be found in a select few places today—most famously at Hamelin Pool in Shark Bay, Western Australia, where the world's biggest colony was discovered in 1956. These modern examples act as excellent analogues, providing an opportunity to under-stand how their fossil counterparts formed and functioned and giving us a little glimpse of what Earth may have looked like then.

Given that we are dealing with multibillion-year-old bacteria, it may seem impossible to understand their reproductive behavior, and you might be wondering, "Hey, where are all those dinosaurs having sex?" Don't worry, we will get to that. Conversely, by comparing these early organ-isms with their modern stromatolite and bacteria counterparts, it is evi-dent that these ancient, single-celled life-forms all reproduced asexually. Further, they reproduced in a very basic way, by simply dividing. This allowed them to multiply in great numbers, take over the planet, and ulti-mately create a world of slime. Paints quite a picture, huh? It may not be glamorous or sophisticated, but in this alien world, these mass sheets of microscopic bacteria provide the earliest evidence for the reproductive behaviors of life on Earth.

Our first potential evidence for sexual reproduction in a multicellular organism comes from something that is surprisingly familiar: algae. This time, it actually is true algae. From washing up on beaches to turning aquariums green, numerous forms of algae are found worldwide. Many reproduce asexually, some sexually, and others can do both, but the oldest fossil evidence comes from a billion-year-old species of red algae called *Bangiomorpha pubescens*. Collected on Somerset Island in Nunavut, Canada, studies of its fossils and their associated spores, compared with the living red alga known as *Bangia*, to which it is very similar, show that individuals of this species vary in form and reflect different reproductive types. That means they may have been capable of both asexual and sexual reproduction, which was accomplished by the release of spores and gametes (their sperm) into the water, just like living red algae. Clearly, for the first few billion years, reproduction was on the smallest of scales—that is, in the smallest of organisms. However, everything would change with a discovery in the Australian outback in 1946.

While exploring the Ediacara Hills, a remote area of the Flinders Ranges north of Adelaide, Australia, the geologist Reg Sprigg discovered something spectacular. He found the fossilized impressions of what appeared to be soft-bodied organisms, somewhat resembling jellyfish and worms. These fossils were presumed to be Cambrian in age because Precambrian rocks were thought to be devoid of complex multicellular life-forms that were visible without the aid of a microscope. This viewpoint, however, was quite wrong. The funny thing is that the remains of such animals had been described long before, even from the mid-1800s in England, but those opinions were dismissed. It was not until 1957 that the chance find of a frond-like fossil called *Charnia*, collected from indisputable Precambrian rocks in Leicestershire, England, eventually led to the realization that the Australian material was indeed Precambrian in age.

Nevertheless, Sprigg's discovery, the first diverse assemblage of exceptionally preserved Precambrian fossils ever found, opened up new possibilities. While Precambrian rocks were once thought to be devoid of life, people began to search for fossils in them, and it paid off. Today, there are several sites around the world that have yielded remarkable Ediacaran organisms. In fact, the illustrator of this book, Bob Nicholls, was part of a team who visited one such site, known as Mistaken Point in Newfoundland, Canada, to study and reconstruct an incredible fossil bed full of about a dozen exceptional soft-bodied species representing some of the oldest animals on Earth.

All of these sites come under the umbrella of the Ediacaran Period, named for the Ediacara Hills, which extends from about 635 to 538.8 million years ago. To describe this geological time frame by comparing it to a group that we are all familiar with, dinosaurs did not exist for another three hundred million years. Yup, these fossils are that old.

Ediacaran fossils represent the earliest multicellular communities known, of strange soft-bodied creatures that made their lives on the seabed. With the appearance of numerous new organisms, albeit still quite basic in form and many still puzzling paleontologists, came newly evolved methods of reproduction for scientists to ponder, too. One such example is the most abundant fossil to be found near Ediacara; it was formally named in 2008, more than sixty years after Sprigg's original discovery.

This fossil is called *Funisia dorothea*, the genus name meaning "rope" in Latin after the rope-like structure of the fossil and the species name in honor of the mother of Mary Droser, one of the paleontologists who formally described the remains. As a result, *Funisia* fossils have come to be known as "Dorothy's Rope." Droser and her team(s) clearly have fun when naming new species, such as the Ediacaran *Obamus*, which was named in honor of the former U.S. president Barack Obama, specifically for his passion for science and the apparent likeness of the shape of the fossil to his ears.

Something of a head-scratcher for paleontologists, ropey *Funisia* were upright, tubular creatures, tightly packed in groups and anchored to the sandy seabed by a rootlike holdfast, the same thing that anchors seaweed to the seafloor. Hence, they were sessile (unable to move). Although nobody

FIGURE 1.1. Clusters of *Funisia dorothea*. (A) Multiple associated tubes and (B) numerous attachment points clustered together on the seabed.

(Courtesy of Mary Droser)

has been able to decipher exactly what type of creature *Funisia* was, with suggestions that it might have been a type of sponge or cnidarian (the phylum that includes corals and jellyfish), one thing we can say is that it very likely reproduced sexually. We know this because we have the by-product of sex—young *Funisia*, called "spats."

Funisia are found in clusters of similarly sized individuals representing the same growth stages, the largest of which are around 30 centimeters long. Dense clusters of *Funisia* may contain more than a thousand individuals per square meter. Because all of these groups are of the same size, it strongly suggests that *Funisia* released eggs and sperm into the water synchronously (at the same time), perhaps as part of a timed event akin to how many modern-day corals reproduce. This is known as a spatfall, where the fertilized eggs rooted into the seafloor and grew together, resulting in spats (*Funisia*) of the same size and age.

Ropelike, enigmatic creatures that formed spats might not seem sexy, for want of a better word, but at around 565 million years old, this is among the oldest evidence we have for sexual reproduction in an organism other than bacteria or algae. In fact, the initial study of *Funisia* fossils also found that they used a form of asexual reproduction called budding, where a new individual develops as part of an outgrowth from the parent. Evidence showing that *Funisia* may have reproduced in such ways, using similar reproductive strategies that are common and widespread today among many marine invertebrates, showcases how even this early, fundamentally simple ecosystem already revealed signs of complexity.

Rather befittingly, Droser's reasoning for naming the fossil after her mom was simple: "She's come with me on digs and done all the cooking and taken care of the kids. It seemed the right thing to do." I, for one, wholeheartedly applaud the choice of name, not only for recognizing the support from her mother but for the very fact that these *Funisia* fossils show evidence of reproduction. In some ways, they are symbolic of Droser and her mother and of animals starting their journey of life.

FIGURE 1.2. (Overleaf). *Three Generations Tall*

In the cold Ediacaran waters, a lone *Spriggina floundersi* slowly makes its way between the strands of three generations of *Funisia dorothea*.

MEGA MILLIPEDE MATING

Spiders, scorpions, ants, wasps, crickets, crabs, ticks. For some of you, I suspect reading those names has just sent a chill down your spine or made you itch. All of these so-called "creepy crawlies" are examples of arthropods, of which over a million living species have been named and described. Arthropods were the earliest animals to walk on land and become dominant over other types.

Myriapods, the group that includes millipedes and centipedes, represent the oldest definite examples of terrestrial arthropods. Based on rare body fossils and various tracks, evidence shows that they lived on land some 440–425 million years ago. The word *millipede* literally means "a thousand feet," and myriapods represent the leggiest animals on Earth.

I remember when I held a millipede for the first time as part of an animal exhibit that was in my hometown of Doncaster. As it moved, it wrapped around my arm, and when I gently removed it, the feeling of its legs coming away from my skin was akin to Velcro. That strange sensation stuck with me. As a young lad growing up in this old coal-mining town, I knew that fossils had been found here and were part of the famous Carboniferous coal measures, being around 310 million years old. Such fossils were found deep inside underground mines, but as far as I was aware you could no longer find them as all the coal mines and pits had long since closed.

When I began volunteering at my local museum, one of my intentions was to visit local sites that may potentially yield fossils. I was keen to visit old spoil tip locations where waste material was discarded. Lo and behold, in 2012, together with a couple of friends, one an archaeologist and the other a local fossil collector, we struck Carboniferous gold—fossil plants. Before we knew it, we had collected more than thirty different species of plants, several horseshoe crabs, fish remains, insects, and even a beautifully preserved shark egg case, which I was thrilled to find on a return visit (my fellow paleontologist and friend, Jason Sherburn, was present on that day and fondly remembers the huge grin that I pulled); it was the first such specimen ever found in this part of the UK. My team and I went on to describe this site in 2014 in a scientific journal. The great serendipity of the site's location is that it is a ten-minute drive from the house I grew up in! Literally, a fabulous fossil site was right on my doorstep and I had no idea.

In practically any museum display depicting a scene out of the Carboniferous, you can expect to be introduced to an ancient swampy coal forest covered in enormous 50-meter-tall scale trees such as *Lepidodendron*, pigeon-sized dragonflies, and early amphibians crawling out of the water. However, one animal that often steals the show in these reconstructions is a giant millipede.

These giant millipedes belong to an ancient, successful (but now extinct) family called arthropleurids, whose fossils have been found in Europe, North America, and Asia. When I say "giant millipede," I really do mean giant. They are the largest arthropods and invertebrates to live on land, perhaps even *the* largest arthropods in Earth's history. One exciting fossil of the most famous of these mega millipedes, called *Arthropleura*, was discovered in January 2018 at Howick Bay near Alnwick in Northumberland, northern England, and it was described in 2021 as the largest *Arthropleura* fossil known thus far. Although incomplete, the three-dimensional fossil comprised a large portion of the exoskeleton that reliable estimates suggest the living individual would have been about 55 centimeters wide and up to 2.6 meters long, it would have weighed at least 50 kilograms (by comparison, the largest living species grows to about 40 centimeters long). Truly, a giant.

The first *Arthropleura* fossils were recorded over 170 years ago. Today, their fossils are known from the Early Carboniferous to Early Permian, although large examples are highly incomplete, with the odd piece here or there. If you are lucky, there might be a partial specimen or something a little more complete, such as the smaller known juveniles, but they are rare. Among the most curious fossils attributed to *Arthropleura* and its kin are the many tracks that have been recognized. These tracks always consist of two parallel rows of closely spaced footprints made by the legs on either side of the body and, although no *Arthropleura* has yet to be found dead in its tracks, the footprints were undisputedly created by a large millipede.

To help easily identify and compare these types of tracks, they were given the scientific name *Diplichnites cuithensis*. Numerous specimens representing single or multiple trackways have been found at various locations in Europe and North America, with perhaps some of the more famous specimens being found in Scotland. The large size of the tracks, reaching up to 50 centimeters in width in some cases, along with their shape and comparisons with tracks made by living millipedes, provide further compelling evidence that they were created by giant millipedes. This is further supported by the fact that no other animal of this size or with a similar shape has been discovered in rocks of the same age.

Rather befittingly, the *Arthropleura* specimen studied in 2021 comes from a geological layer that is directly contemporaneous with examples of large *D. cuithensis* tracks. This represented the first instance of a giant arthropleurid body fossil from within the same regional geological sequence as the tracks, meaning they are from the same age and environment, and that is inherently rare. Rarer still is an unusual *Arthropleura* track that strongly indicates one pair of these mighty millipedes got down to business.

Of course, sex is a major part of life. I suspect, though, that you have never given too much thought to millipede mating matters, unless you are a diplopodologist (a zoologist who studies millipedes, not the dinosaur *Diplodocus*). That said, the chance of finding and identifying a giant millipede mating track would appear unlikely. Yet, as with many fossil finds, this was another one discovered by chance.

A study published in 2005 focused on an intriguing and unique fossil trackway found on a beach near St. Andrews in Fife, Scotland. Dating to the Early Carboniferous, roughly 330 million years ago, this track was made by a huge, six-legged sea scorpion (called a eurypterid). The study was undertaken by Martin Whyte, a lecturer and well-respected paleontologist known in paleontology circles for his extensive research, especially on Jurassic dinosaur footprints from the Yorkshire Coast in England. It was Martin and another colleague, Mike Romano, who reported the first stegosaur footprints in the world, which were found in Yorkshire. In 2010, Whyte, David Williams, and Dee Edwards returned to the beach in Fife to reexamine the eurypterid track and to create a plaster mold of it, to ensure that an additional record exists. It was during this trip that the team spotted nearby three closely associated arthropleurid trackways belonging to *Diplichnites cuithensis*. The large sandstone block containing the tracks had clearly been exposed for some time, showing signs of erosion and being partly encrusted with modern barnacles and limpets.

FIGURE 1.3. (A) *Diplichnites cuithensis* track with zoologist Sally-Ann Spence for scale. (B) A modern pair of mating millipedes in parallel position.

([A] Courtesy of Sally-Ann Spence; [B] courtesy of Bojan Ilić and Zvezdana Jovanović)

FIGURE 1.4 (Opposite). *Lovers Entwined*

Two enormous millipedes, *Arthropleura* sp., an amorous male and a female, meet on a Carboniferous floodplain and begin their enchanting courtship.

Most curiously, unlike any previously known examples of these types of tracks, they exhibited signs of interaction between the trackmakers walking in the same general direction. The longest and best-preserved trackway, which measures an impressive 21.5 meters, is directly associated with another trackway measuring 7.3 meters. This smaller track runs alongside, crosses, and even slightly overlaps the large track, resulting in a short section of the larger track having a third line of footprints, which suggests one of two things. Either two arthropleurids were walking one behind the other, with their footprints interlocking and matching exactly on one side for some distance, or one of the arthropleurids was partially mounted on the other, thus producing the three parallel lines.

The first interpretation seems highly unlikely, as it would mean the second millipede would have had to place its numerous feet perfectly into each of the footprints of the other. Moreover, there is a noticeable difference in the width of the two tracks. The larger track is 40 centimeters wide, and the smaller track is narrower, only 35 centimeters wide, meaning that the millipedes were slightly different in size. That makes it even more unlikely that the millipedes would have matched their tracks perfectly, since the distance between their feet (and footprint stride) would have differed. The only plausible explanation is that the slightly smaller millipede was partly riding on the back of the other. This is further supported by the fact that the closely paired grooves of both footprints are deeper on one side of the track, suggesting a greater weight on that side.

Okay, so why? We can rule out an attack from behind because these millipedes dined on decaying matter and were not predatory; at least that is the most recent understanding of their diet. This leaves us with the most reasonable and logical reason for one millipede mounting the other: sex.

The structure of the track suggests that the millipedes were partly intertwined toward the front of their bodies, with some indication that the smaller individual was lying on the left side of the larger individual. Being connected at the front makes sense because both male and female modern millipedes are what we call progoneate, meaning their genital opening is situated in the anterior part of the body. Interpreting this further, at one point during the interaction the larger trackway ends abruptly, with no footprints present, and then continues in a slightly different direction. This deviation suggests they must have reared up further off the ground to adjust their position while mating before placing their feet down again and moving on.

This interpretation of a sexual encounter can be corroborated further by the fact that in many modern millipedes, the male approaches from behind and stimulates the female by climbing on her back before the two get it on. Connecting their anterior parts, most male millipedes then use their legs (called gonopods) to pass on a sperm package to the female. Mating may last anywhere between an hour to two days, which may have been the case for our arthropleurid pair.

I guess we will never know the ins and outs of this intimate moment, or how long it lasted, but this unique fossil gives us a glimpse into the sexual behaviors of these giant arthropods, one of which we may wish to ponder. Or not.

GIANT *T. REX* PENIS

Admit it, you saw this title in the contents list and quickly came here looking for a giant *Tyrannosaurus rex* penis in all its glory. I can understand because it would genuinely be intriguing to see, plus we paleontologists know that finding a *T. rex* penis would be a big deal and would no doubt be a solid feature in the news all around the world. But I am sorry to say that we have yet to find that elusive *T. rex* phallus. Apologies for getting your hopes up. This does, however, lead us nicely into the exciting realm of dinosaur sexy time, with one pretty incredible fossil that is exceptionally preserved, taking the "bare bones" of fossils to another level.

The incredible fossil in question is an example of the Labrador-sized ceratopsian, *Psittacosaurus*, a member of the typically horned, beaked herbivorous group of dinosaurs. Before going further, it is worth a comment on the pronunciation of this dinosaur's name. I have heard people pronounce it in all sorts of ways, including "pistachiosaurus." But the easiest way to say the name is to omit the "P" (like in pterodactyl) so that it sounds like "sit-taco." In this case, maybe we could refer to this dinosaur as a tacosaur. I am sure it ended up being food for plenty of carnivores.

This primitive bipedal cousin of *Triceratops* is by far one of the best-known early ceratopsians, with fossils found in museums around the world. One of today's most famous specimens comes from the Jehol Biota of China's Liaoning Province and is roughly 125 million years old; it is referred to in paleo circles as the Frankfurt *Psittacosaurus* after the museum where it is on display, the Senckenberg Natural History Museum in Frankfurt, Germany. What makes this one stand out from the busy *Psittacosaurus* crowd is that it represents the best preserved nonavian (bird) dinosaur yet described, with scaly skin and remarkable soft tissues; it even contains the only known umbilical scar from a "belly button" in a dinosaur. Oh, and it preserves the world's first dinosaur cloaca! That is a big deal.

But what is a cloaca, you ask? Perhaps the best way to explain this is to ask whether you can remember the first time you saw a bird sitting on a nest of eggs? I know I can. It was part of a school trip to a local farm when I was about five years old where I saw some hens sitting on their eggs. Then I discovered that the eggs apparently come out of a chicken's butt, and we eat them. I wonder what people thought of the first person to do this.

Anyway, as you have probably guessed, the opening for the chicken's "butt" is the cloaca (the Latin word for "sewer"), a multipurpose hole used for both reproduction and excretion. The eggs do not really come out of the butt per se because they are made in the ovary and come out of the tubelike oviduct rather than the intestine, but they do come out of the cloaca, which contains a single orifice often called the vent.

Now that we have that out of the way, it might seem a little odd, but paleontologists have been searching for confirmation of the fabled dinosaur cloaca for some time. By looking at living dinosaurs (= birds) and their closest living relatives, the crocodilians, it was inferred that the same sex parts would also have been present in their extinct relatives. As such, all male and female birds and crocs have a cloaca, which strongly suggested that dinosaurs had a cloaca as well. This is why the *Psittacosaurus* cloaca was so pleasing to identify.

This is where I should probably hand over this part of the story to the illustrator of this book, Bob Nicholls, who was part of the team who studied and described the cloaca; I will pass the writing to him in just a moment. He joined forces with the University of Bristol paleontologist Jakob Vinther and the University of Massachusetts Amherst biologist Diane Kelly, who together described the cloaca in 2021 by comparing it with similar openings in many living animals. As this fossil bears such exquisitely preserved soft tissues, when looking at the specimen it can be quite the challenge for your eyes to wander to the nether regions because the fossil is so visually spectacular. Although it is where you would expect it to be, it can be hard to spot the cloaca positioned slightly behind the legs on the underside and back from the ischial callosity (or "sitting pad").

To help in understanding more about the shape of the cloaca, this direct quote from Vinther in an interview with CNN summarizes it perfectly: "It is very unique. Most cloacas form a kind of slit. Sometimes it's a vertical split, sometimes it's a smiley face, sometimes it's a sour face. This thing has a V-shaped structure with a pair of nice flaring lips and there's not a living group of animals that have morphology like that. It is somewhat similar to a crocodile's but still unique." I grin at the fact that he referred to this as having a "pair of nice flaring lips." As amusing as this might be, Vinther and the team really emphasize the point of just how unique this cloaca is.

The opening (vent) has a clear difference in scale anatomy and pigmentation, so much so that it is distinct from the adjacent body regions. Although

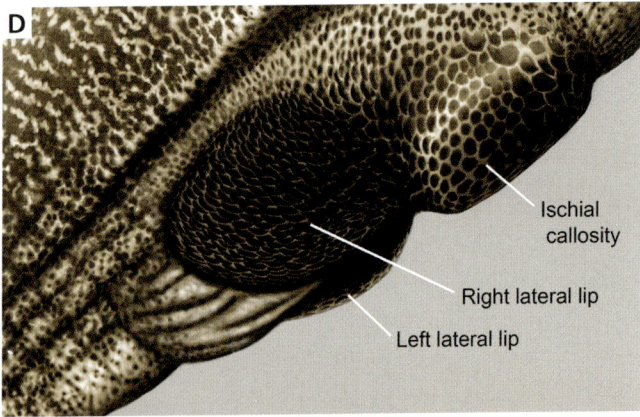

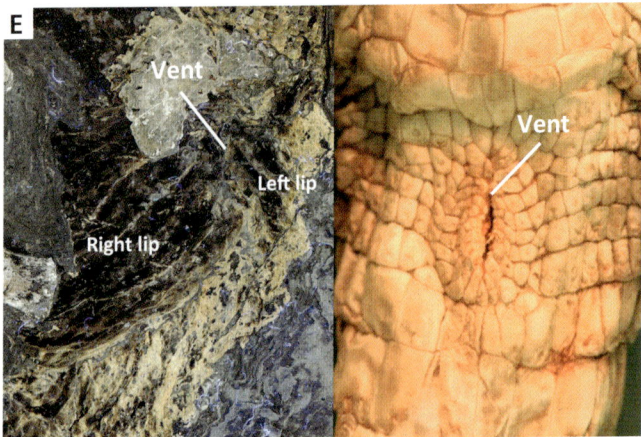

the cloaca does not provide details about the sex of this animal (no penis or clitoris could be located), the cloaca clearly stood out from the rest of the skin and suggests possible roles for visual signaling, meaning that it was probably used to show off and attract other *Psittacosaurus*; cloacal signaling does occasionally occur in birds today. Additionally, it was probably also used for olfactory signaling because the lateral swellings of the vent are in a similar position to the paracloacal musk glands in crocs, which produce a distinctive odor that is released during social displays by both males and females; that might suggest Taco did the same. As a bit of a bonus for this fossil, it is worth adding that a cream-colored mass located immediately inside the cloacal opening almost certainly represents a coprolite, a fossil poop.

Not only was Nicholls involved with the science, but he was also tasked with bringing this fabulous *Psittacosaurus* and its curious cloaca to life with a 3D scale model. In order to do that, he examined the specimen in detail, spent days illustrating the specimen in person, snapped numerous photos, and studied the cloacae of modern animals. Nicholls can be credited as the very first person to scientifically and accurately reconstruct a dinosaur butthole. That is definitely something to be proud of. I asked him if he ever thought he'd one day reconstruct the first dinosaur butt and this was his reply:

> No, certainly not! Reconstructing the life appearance of extinct animals is my job, so I have spent more time thinking about dinosaur bottoms than just about . . . well, everybody. They are just another part of a dinosaur's anatomy that needs illustrating after all, but I never thought I would be the first person in the world to reconstruct one accurately. Being one of the three authors, with Jakob and Diane, to be the first to formally describe the *Psittacosaurus* cloaca in a scientific paper was also a huge thrill. The media coverage around the world was extraordinary! It was wonderful to see so many people getting excited about a dinosaur butthole.

FIGURE 1.5. (Opposite). (A) The exquisitely preserved *Psittacosaurus* and its colorful cloaca. (B) A closer look at the cloacal vent region. (C) A close-up of the cloaca's "nice flaring lips" (right side). (D) 3D reconstruction of the cloacal region seen in (B). (E) *Psittacosaurus* cloaca compared with a freshwater crocodile cloaca.

([A-D] Courtesy of Bob Nicholls; [E] courtesy of Mike Pittman, Thomas G. Kaye, and Phil Bell)

As an excellent side note, Nicholls and Vinther were already familiar with this fossil (and had noted the presence of the cloaca) since they were the leading figures in an earlier study that described the preserved color pigments in this *Psittacosaurus*, which suggested countershading camouflage. As they discussed in their study, this type of countershading uses a dark-to-light gradient from back to belly to counter the light-to-dark gradient created by illumination. As a result, the body appears flatter and less conspicuous, and this pattern would have been ideal in a closed forest habitat where we know *Psittacosaurus* lived. Based on all of this, Nicholls's 3D reconstruction has been regarded as the "most accurate depiction of a dinosaur ever created."

Dinosaur sex. A preserved cloaca. Discussions of dinosaur penises. The press had a field day with this research and rightly so. How often are we allowed to talk about these types of cool albeit weird aspects of paleontological research? Some humorous headlines I could find included these gems: "1st Preserved Dinosaur Butthole Is 'Perfect' and 'Unique,' Paleontologist Says" (Live Science); "Finally in 3-D: A Dinosaur's All-Purpose Orifice" (*New York Times*); and "This Fossil Reveals How Dinosaurs Peed, Pooped and Had Sex" (CNN). However, by a long shot, the most hilarious coverage of the study was featured on *The Late Show with Stephen Colbert*, where the host not only discussed the research and showed Nicholls's artwork but also went on to say that "Gwyneth Paltrow has released this candle that smells like a *Psittacosaurus* cloaca." Of course, that candle was not released, but maybe it should be.

Regarding how *Psittacosaurus* actually did the deed is another question. For instance, in modern crocs the penis remains hidden inside the male's cloaca and pops out during sex, whereas most male birds lack a penis but have sex (and exchange sperm) through what is known as a "cloacal kiss," where the male and female cloacae come together and touch. It is hard

FIGURE 1.6. (Overleaf). *Securing the Species*

Following a successful courtship, which included dancing and cloaca sniffing, these two "tacosaurs" begin the essential task of producing more *Psittacosaurus* dinosaurs. Dino sex as it happened.

to say for sure whether a male *Psittacosaurus* had a penis, though further studies of this fossil cloaca have found that the opening was indeed more like that of a croc and, therefore, it almost certainly had a similar internal anatomy, suggesting it would probably have had a penis.

Thinking about it, I should probably have given some parental advisory warning about this naked fossil, but it has been preserved for 125 million years and is openly on public display in all its glory. So, I am sure you will agree just how cool it is and hope that one day you might visit this fossil and stare into the abyss of its cloaca, pondering the sex lives of dinosaurs. If you do, then I have done my job.

EIGHTEEN AND COUNTING

In a while crocodile, see you later alligator, shed a tear for our extinct choristodere. Not quite the same ring to it, I know, but if you saw a choristodere alive today you might think it was just another crocodilian. You would be wrong.

Choristoderes are a distinct group of reptiles whose fossil record extends back into the Middle Jurassic and perhaps even earlier. They were a successful group that barely survived the catastrophic events of the asteroid impact at the end of the Cretaceous, but the group's final members disappeared a little over 10 million years ago.

Although most choristoderes looked lizard-like, some species appeared more croccy, even though they lacked a body covered in osteoderms (the armored bony plates in crocodiles). It is probably worth pointing out that perhaps the most famous of all choristoderes is the Late Cretaceous–Early Paleogene *Champsosaurus* from western North America, whose name literally means "crocodile lizard" in Greek. Named in 1876 by Edward Cope (of the Bone Wars fame), *Champsosaurus* is a name you might be familiar with as it is often included in dinosaur and prehistoric animal fact file–type books, where hundreds of prehistoric creatures are profiled. With its familiar croccy look it is easily missed in favor of more impressive-looking animals, but it is a real champ for this group of extinct reptiles. In any case, from here on out, let us call them choristos, as it is easier to say.

Our focus is on a Cretaceous choristo from China. It was during the Early Cretaceous that choristos hit their peak in diversity; several long- and short-snouted new species appeared, thus representing an important time in their evolutionary journey. Most of these Cretaceous species come from Asia, such as *Hyphalosaurus baitaigouensis*. Yes, I struggled to pronounce that name, too. This unusually long-necked reptile had a tiny head and an especially long tail, and adults grew to about 1 meter in length. Thousands of *Hyphalosaurus* fossils have been collected from the famous Jehol Biota fossil beds of Liaoning Province, which are around 120–125 million years old and date to the Early Cretaceous, but we are interested in two exceptional individuals who were pregnant.

Reading the word "pregnant" and knowing this is a prehistoric reptile, you may immediately think of those exceptional fossils of ichthyosaurs, the mighty marine reptiles, which have been found carrying unborn fetuses. At least, *I* would think this, given that I have spent most of my career studying ichthyosaurs and describing several pregnant (gravid) specimens. In fact, as far as strange fossil titles go, ichthyosaurs hold the title of the reptile group with the most pregnant fossils found, with more than one hundred gravid females known (such as "Fiona" mentioned in the preface of this book). To avoid any misunderstanding, ichthyosaurs were viviparous reptiles, meaning they gave birth to live young. For that matter, I should point out that we also have one fossil of a pregnant plesiosaur.

With that said, it is worth mentioning some obscure but exceptional examples of other viviparous marine reptiles that you may not be familiar with. Perhaps the most exciting, at least from a paleontologist's perspective, was the 2017 discovery of a specimen of the bizarre, long-necked marine reptile called *Dinocephalosaurus*. Found in southwest China, this fossil dates to about 245 million years ago, from the Middle Triassic, and was found with a single embryo inside. Paleontologists were particularly excited by this discovery because it represents the earliest evidence yet of live birth in the same wider group of animals (called archosauromorphs) thought only to lay eggs, including crocs and birds and their extinct relatives, such as the nonavian dinosaurs and pterosaurs.

Another remarkable find from Triassic China was the discovery of two gravid examples of the small (20–30 centimeters long) marine reptile called *Keichousaurus*. One contained four embryos, and the other contained at least six. This was a significant find as *Keichousaurus* is a primitive member of a large group of ancient marine reptiles called sauropterygians, which includes plesiosaurs, among others. This was the first unequivocal evidence of the reproductive mode (live birth) in a sauropterygian. Even the long-snouted, toothy, but tiny mesosaurs, representing the earliest reptiles to adapt to an aquatic lifestyle in the Early Permian, probably gave birth to live young, too.

It is also worth highlighting that at least one mosasaur precursor has been found with four embryos. The mosasaurs are a famous group of marine lizards from the Late Cretaceous, and this early forerunner found in

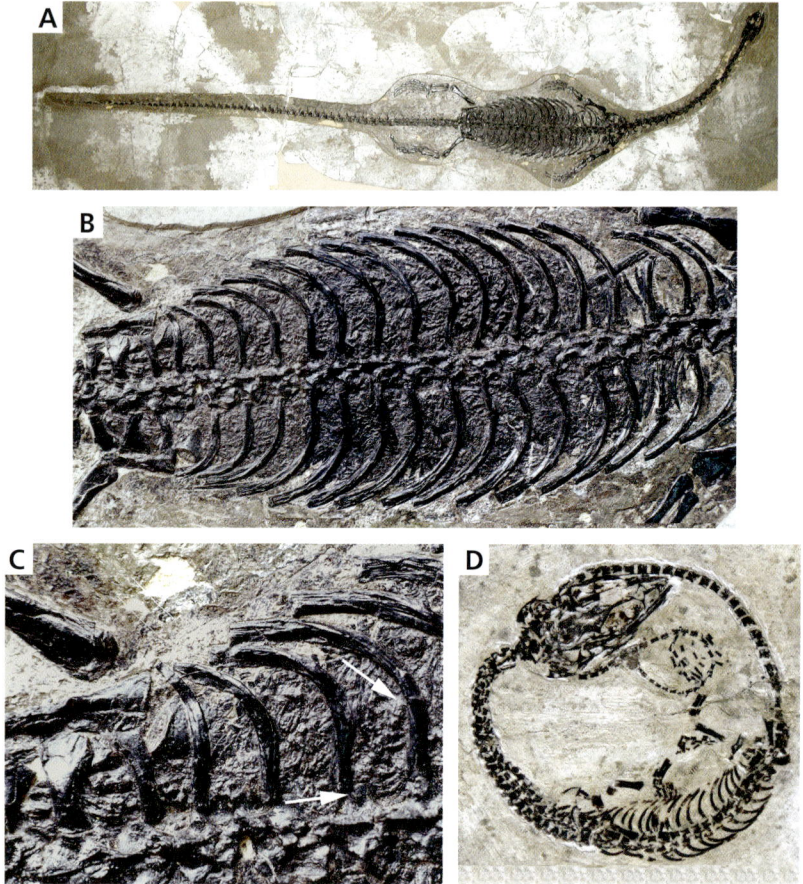

FIGURE 1.7. (A) The gravid choristodere, *Hyphalosaurus baitaigouensis*, with eighteen embryos packed inside. (B) Trunk region of the adult with densely packed embryos. (C) A close-up of the posterior portion of the adult showing numerous embryos, including a distinct line of vertebrae (arrow). (D) A small, free-living individual compared to the embryonic remains.

(Courtesy of Xiao-Chun Wu)

FIGURE 1.8. (Opposite). *The Freshwater Litter*

In a shallow river, beneath a warm sky, a mother choristodere (*Hyphalosaurus baitaigouensis*) gives birth to eighteen lively young.

Slovenia, called *Carsosaurus*, belongs to a primitive group known as aigia-losaurs. *Carsosaurus* was possibly semiaquatic and gave birth to live young but was not adapted to live entirely in the marine world. However, we know that later members of the mosasaur group became fully aquatic. We have further support for this in the form of neonate mosasaurs found in what would have been open ocean environments.

All of this is important for our choristos because ichthyosaurs and plesiosaurs gave birth to live young out at sea, whereas choristos did not. A crucial part of the choristo story is that—as far as we know—they never adapted to live in the seas or oceans but only inhabited freshwater environments such as rivers and lakes. A few plesiosaurs and mosasaurs also inhabited those locations, but they evolved from marine ancestors, whereas *Hyphalosaurus* and friends did not.

Of all the choristos known, *Hyphalosaurus* was among the most aquatically adapted. Fossils are commonly found inside rocks that were formed in medium-depth to deepwater lake deposits. Despite the aquatic adaptations, the anatomy of its limbs suggest it may still have been capable of walking on land, although its long neck and long tail probably meant that it looked a little clumsy when moving around.

Given the aquatic lifestyle of *Hyphalosaurus*, it could be surmised that it did everything in the water, including giving birth. Confirmation of this mode of reproduction came in the summer of 2007 when an exquisite, 80-centimeter-long, almost complete individual containing an astonishing eighteen embryos was discovered. This is the sole record of live birth in a prehistoric freshwater reptile and the most embryos found inside a fossil reptile, at least to my knowledge. Quite the title for little *Hyphalosaurus*.

The embryos filled the body cavity and were arranged in pairs, most of them in a curled position, likely snuggled up in their egg sacs. By comparing the tiny embryonic skeletons with several free-living individuals of about the same size, along with the more straightened body postures of the pair closest to the pelvis, it became clear that the expectant mother had probably reached the point of, or was very close to, parturition (giving birth). In the line of embryos, the last one located on the left side is positioned further away from the others and had rotated in such a way that the head pointed toward the pelvis, contrary to all the others. Rather sadly, this abnormality suggests a potential complication occurred that may have led to the death of the mother and all of her young. A great tragedy.

Interestingly, when the *H. baitaigouensis* species was formally described in 2004, the key (holotype) skeleton was said to be associated with eleven eggs containing embryos. Further examination of this specimen found there to be fourteen in total: twelve preserved outside of the body and two inside; the external ones were most probably ejected from the mother's body after death due to a buildup of gases.

You might be wondering why we have one pregnant specimen with no sign of eggs and one with eggs. In various living squamates (lizards, snakes, and slow worms), during gestation, there are intermediate stages between oviparity (with eggs) and viviparity (live birth). This form of reproduction is traditionally called "ovoviviparity," where the eggs develop and hatch inside the mother's body before they are delivered, but it is still correct to refer to this as a form of live-bearing reproduction or live birth.

The individual with eighteen embryos was clearly very close to giving birth. Each embryo was at an advanced stage of development, around the same size as free-living individuals of the same species. Plus, as above, the last two embryos had a straightened posture and were positioned in the pelvic region. By comparison, the embryos in the other specimen were in an earlier developmental stage, and each egg lacked any trace of the shells. It seems likely that the loss of the shell occurred during the transitional period from oviparity to viviparity, which is common among living lizards.

However, as a bit of a curveball, two isolated and *egg*ceptionally rare soft-shelled eggs were also found—one contained a *Hyphalosaurus* embryo and the other was associated with a partially hatched neonate that was about 6 centimeters long (tiny compared with the adults, which reached 1 meter). These flexible eggs had poorly mineralized eggshells, suggesting that *Hyphalosaurus* may have given birth to live young *and* laid soft-shelled eggs. This is called bimodal reproduction, where said animal is capable of both live birth and egg laying. The living three-toed skink, a small lizard endemic to Australia, is one of the most famous species capable of bimodal reproduction. Quite extraordinarily, just in 2020, one of these skinks was found to have laid three eggs and later gave birth to a fully developed individual, all from the same clutch! This represented the first documented occurrence of such an unusual phenomenon in a vertebrate animal. In any event, the mother pregnant with eighteen embryos confirms live birth in *Hyphalosaurus*.

Although the skeletal features of the neonate match what is currently known for the anatomy of *Hyphalosaurus*, there lies a slim possibility that the two eggs belonged to a very close relative of *Hyphalosaurus* instead. Based on what we currently know, however, findings suggest that some *Hyphalosaurus* probably laid squishy eggs out of the water in closed nests on the edges of lakes and rivers, whereas others gave birth in the water. This latter situation may have been critical to the survival of this little reptile because it could avoid the threat of not only having its offspring eaten but also being eaten itself.

Based on fossils found in the same area, we know that carnivorous dinosaurs (including birds), lizards, and mammals posed a serious threat on land and provided intense competition. The watery transition would essentially have freed *Hyphalosaurus* from the perils of terrestrial predators, which meant there was a lower risk of death and a higher chance the tiny offspring would survive to fight another day.

THE "LIZARD FISH" MOMS

One of the many cool things about being a paleontologist is that you might get to name an entirely new species of prehistoric animal. Most names are traditionally taken from Greek or Latin and may focus on a characteristic feature or features of an individual, such as *Triceratops*, whose name means "three-horned face" in reference to the pair of massive brow horns and the nasal horn. You can get really creative and have some fun with this, too. One of my favorite fossil species names is *Vaderlimulus*, named for a horse-shoe crab with a head shield that sort of resembles Darth Vader's helmet. In most of the species I've named so far, my preference has been to name species after prominent individuals, such as *Ichthyosaurus anningae*, a new species of ichthyosaur I named in 2015 in honor of Mary Anning.

Sometimes the name bestowed on a prehistoric animal might be a little confusing; for example, the word *ichthyosaur* means "fish lizard" in Greek, yet ichthyosaurs were neither fish nor lizards. That brings us to the "lizard fish." This name might conjure up some strange imagery of an unusual, prehistoric lizard-like fish, but the reality is that "lizard fish" comes from the naming of a Triassic fish called *Saurichthys* (Greek for "lizard fish"). This name was introduced in 1834, sixteen years after the ichthyosaur was named. Clearly, fish lizards and lizard fish had captured the attention of pioneering paleontologists.

With its elongated body and long, narrow jaws, *Saurichthys* superficially resembles a modern garfish and belongs to a group of bony fish known as ray-finned fishes, named for their fan-like fins made up of spines called rays. Chances are, if I asked you to name a ray-finned fish alive today you could guess and you'd almost certainly get it right because there are more than thirty thousand living species, which make up around 95 percent of all living fish. For instance, salmon, seahorses, and goldfish are all ray-finned fishes whose ancestry stretches back more than 400 million years.

Of all ray-finned fishes, live birth has evolved independently at least twelve times among the various groups. For example, all members of the seahorse family, Syngnathidae, give birth to live young, although they are a pretty radical example because it is the male who becomes pregnant, incubates the developing embryos, and gives birth. In the entire fossil record

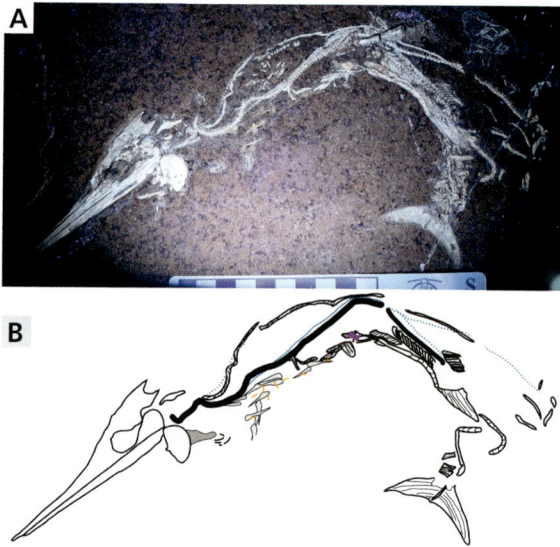

FIGURE 1.9. Two pregnant *Saurichthys*. (A) A complete specimen of *S. curionii* containing multiple embryos, photographed under UVA light. (B) Interpretive drawing of the same specimen; gray triangles represent embryonic skulls and orange lines indicate an embryonic notochord.

of ray-finned fishes, the only known evidence we have of live birth comes from two species of *Saurichthys*, called *S. curionii* and *S. macrocephalus*. This highlights the extreme rarity of such fossils, especially when we consider the great diversity of ray-finned fishes today and the numerous extinct species we have in the fossil record.

The *Saurichthys* fossils come from a famous area in Monte San Giorgio, Switzerland, which is designated as a UNESCO World Heritage Site due to the significant paleontological discoveries made there, primarily because of the exceptional fossil preservation. Hundreds of *Saurichthys* specimens have been unearthed, some including evidence of muscle tissues and even intestine traces; they date to the Middle Triassic, approximately 240 million years ago.

With such exceptional preservation and numerous specimens at hand, one team focused on attempting to understand the ontogeny (growth) and reproductive biology of *Saurichthys* by studying fossils of embryos through juveniles to subadults and adults. As you can imagine, having a full set of

FIGURE 1.9. (*Continued*). (C) A close-up of the main group of embryos. (D) One of the embryonic skulls with associated soft tissues.

(Courtesy of Erin Maxwell)

ontogenetic stages for an extinct vertebrate like this is quite rare, but it provides an excellent opportunity to understand how an individual changes with age. This is important for paleontologists because the differences between a very young individual and a fully mature individual may give the impression that they are different species, so having those age gaps filled in can ensure mistakes are not made. Male and female *Saurichthys* can also be distinguished based on the presence of a mineralized copulatory organ (i.e., its fun bits), and females are larger than males.

The team, led by my friend Erin Maxwell from the State Museum of Natural History in Stuttgart, Germany, carefully examined specimens for embryonic remains but quickly ran into issues. Distinguishing embryos versus stomach contents (prey items) can be tricky, particularly when the remains are so small, but also because several confirmed cases of cannibalism have been thoroughly recorded in *Saurichthys*. To combat this, Maxwell's team came up with a couple of criteria.

First, in some rare cases, the stomach was well preserved and could clearly be distinguished, so any individual found in the stomach contents or positioned in or around the mouth and throat region was identified as ingested

prey items (food). Any small individuals that were outside of the stomach and positioned below the vertebral column were considered embryos, but multiple individuals of the same size would also suggest embryos because prey would be expected to vary in size. Furthermore, some of these tiny individuals in the abdominal cavity had exceptional three-dimensional soft tissues present, which is not seen in smaller specimens observed in the gastrointestinal contents. That was further evidence for embryos.

The researchers identified eighteen large females with small specimens inside the abdominal cavity, but considering the challenges of identifying embryos versus food, only six were found to be unambiguously gravid. The pregnant females ranged in size from 30–60 centimeters. Each of the six individuals contained many embryos, with one of them comprising at least twenty-four embryos.

Most of the embryos are preserved together in clusters inside the front and middle parts of the abdominal cavities. In two of the specimens, two clumps of embryos are preserved on either side of the body, which may correspond to the left and right oviducts. The most easily identifiable parts of the embryos are the skulls, which range from just 9–16 millimeters long; tiny teeth can also be seen. By comparing the largest embryos with the smallest neonates, the latter with jaw lengths of about 20 millimeters, it was determined that birth must have occurred once the embryos reached approximately 7–12 percent of the adult's body length.

When an animal is pregnant it requires more energy and essential nutrients to help support the growth of its developing young, so finding specimens of *Saurichthys* with both embryos and stomach contents inside is quite special and provided evidence for two distinct types of behavior. Having multiple gravid specimens of *Saurichthys*, with one containing more than twenty-four embryos, these fascinating fossils provide the oldest evidence yet for viviparity in ray-finned fishes, undisputedly one of the most successful groups of vertebrates.

FIGURE 1.10. (Opposite). *Hiding from the Cannibal Crowd*

A large female fish, *Saurichthys macrocephalus*, has sought out a secluded space between some rocks. Away from the passing school of *Saurichthys*, she gives birth to twenty-four healthy juveniles.

2 | Eggs and Babies

After producing their offspring, a few spiders, such as Stegodyphus lineatus, *make the ultimate sacrifice: The spiderlings eat their mother alive. This extreme type of brood care is called matriphagy or suicidal maternal care.*

TETHERED TODDLERS

The fossil record has thrown up so many examples of weird and wonderful creatures that if they were conjured up in Hollywood, people would likely consider them too unrealistic to be taken seriously. I'm thinking of critters like the Cambrian *Opabinia*, which had a long mouthpart ending in a claw and five mushroom-shaped eyes on the ends of stalks. Yeah, pretty weird. There are many other crazy creatures like this in the fossil record. But what are the chances of finding something so unusual that it hints at an entirely unique behavior not previously recorded in any living or extinct critter?

With few (or no) living analogues for comparison, it is no surprise that paleontologists may struggle to understand and often argue among themselves regarding the behaviors of such creatures. That is why we must thank our lucky fossil stars when evidence comes our way.

You might think that the fossil in question comes from some far-flung location, perhaps remote icy Patagonia or the searing hot Sahara Desert, but no. Our fossil comes from the hills of Herefordshire, a rural English county in the Welsh Borders, which has revealed some remarkable fossils over the past thirty years.

This deposit is known as the Herefordshire Lagerstätte, a site of exceptional fossil preservation of animals that were entombed in ancient volcanic ash, and it dates to the Silurian Period, roughly 430 million years ago. Some exciting fossils come from here, not least the world's oldest-known penis, which belongs to a species of ostracod called *Kolymbos sathon*. Yeah, that is a bit of a mouthful, but the soft parts of this tiny ostracod, a type of shelled crustacean, including the penis, are preserved in all its glory. Its name even translates to "swimmer with a large penis."

Considering that we are dealing with a 430-million-year-old arthropod penis, you can begin to understand why this site is known for its exceptional preservation. However, these fossils are preserved in three dimensions as calcite infills inside hardened concretions called nodules, and the only way to study them is to, well, destroy them. What?! I know this is not what you were expecting to read, but the very nature of these fossils means that to unlock their secrets, scientists must very carefully extract the remains through a specialized grinding technique, which removes the tiniest fractions of rock at set intervals. Slice by slice, each newly exposed surface is then recorded as a digital image and stacked to generate an exact 3D virtual fossil and model.

In 2016, a team of paleontologists led by Derek Briggs of Yale University, a renowned specialist in exceptionally preserved fossils like these, described a new type of arthropod from one of the Herefordshire nodules. Making up part of the paleo team were twin brothers, Derek Siveter and David Siveter, from the University of Oxford and the University of Leicester, respectively, who, together with Briggs, have collaborated on studies of the Herefordshire material since the mid-1990s. This new genus and species of spiny arthropod, called *Aquilonifer spinosus*, was described based on the discovery of a single, almost complete fossil with a body length of just under 1 centimeter. However, it was not necessarily the identification of this as a new species that was so exciting (although it genuinely is an unusual creature) but the fact that ten tiny arthropods were found to be tethered to this animal's tergites (the plates on its body) by long individual threads. You can sort of visualize this a bit like how some parents use walking reins or harnesses with their young children or, I guess, to a lesser extent, having your dog on a leash.

This discovery is quite sensational because it suggests a unique type of brood care that has not been recorded anywhere in extant or extinct animals. One of the closest analogues can be found in the living freshwater (marbled) crayfish, where the embryos are tethered to the adult by a stalk. Upon emerging from its egg case, each hatchling remains attached to the parent by a thread until it eventually frees itself.

As you will know if you read *Locked in Time*, we do have some curious cases for brood care in fossil arthropods, including several from half a billion years ago, such as adult-juvenile associations and individuals with eggs attached to their legs. But we also find similar examples of this type of care in modern arthropods, so we assume that prehistoric species probably undertook something similar, unlike what is found in *Aquilonifer*.

The team proclaimed that the ten tethered individuals were juveniles attached to the adult. They even expressed this interpretation in the name, which comes from the tethered individuals sort of resembling kites, but it also (as the authors state) echoes the title of the 2003 novel *The Kite Runner*

by Khaled Hosseini. The name *aquila* means "eagle" or "kite," and *-fer*, the suffix, means "carry." Thus, *Aquilonifer* translates to "kite bearer" and became known in paleo circles as the "kite runner."

At the end of each individual thread, the tiny tots were found inside a shell-like capsule that was shaped like a flattened lemon and ranged in size from ~0.5 to 2 millimeters in length; the thickness of the shell differed, and parts may have been thicker due to soft tissues adhering to the inner surface. How the adult transferred or attached its offspring onto its back is impossible to say, but the researchers suggest that perhaps the long antennae may have been involved, or one parent may have attached the eggs to the other before the threads grew. The little ones had about six pairs of appendages in contrast with fifteen in the adult, which suggests that this animal added segments to its body as it transitioned from a juvenile to an adult, as is known for multiple arthropod groups. The team surmised that the parent is likely to be a female, but that male brood care could not be ruled out as this is also known in arthropods today, such as sea spiders.

As soon as this research was published, it was challenged. Remember what I mentioned above? Paleontologists like to squabble over interpretations such as this, which is sort of how science progresses, at least in a roundabout way. Just over a month following the original publication, a one-page note was published in the same journal offering a different explanation of the apparent brood care. Challenging the original interpretation, this study reasoned that the ten individuals were not juveniles but were probably mites acting as phoronts—these types of animal latch themselves onto a host so that they can get a free travel pass and ride around.

The original team had actually considered the possibility that perhaps the ten individuals might belong to a different arthropod entirely, possibly even a form of parasitic species, although it seemed unlikely given their careful interpretations and the fact that there would be no advantage in

FIGURE 2.1. (Opposite). *Tethered for Safety*

A mother arthropod, *Aquilonifer spinosus*, carries her precious babies across a Silurian ocean floor. She is about to encounter the discarded exoskeleton of another *Aquilonifer*.

such long threads and their attachment point for absorption of nutrients from the host. Plus, this would also have to take into account the molting stage of the parent, where the release of the juveniles would likely have occurred when the adult cast off its old exoskeleton during molting; the team even considered that the adult may have been able to delay its molt until the juveniles were old enough to hatch.

If they were mites, molting would presumably cut them off. Nevertheless, combating this new suggestion of *Aquilonifer*'s kites being mites, the team were quick to review this interpretation (literally on the next page in the journal) and disagreed based on several facts. They further explained that *Aquilonifer* does not appear to have primarily been a swimmer, so it would not have been a good choice for any animal hoping to be taken on a sea voyage.

Subsequent researchers have agreed with the original interpretation that these long threads functioned as a sort of safety line that prevented the hatchlings from being lost. This also meant that the adult could keep its offspring close and protect them from any potential predators—a truly unique and unusual act of behavior that has so far only been documented in this creepy kite runner.

THE CROC GUARDIAN

Picture a *Tyrannosaurus*, a chicken, and a crocodile. Chances are, you will see more immediate similarities between big T and a crocodile. Namely the big toothy grin. But looks are only skin deep because, when studying their anatomy combined with the fossil record, it becomes clear that a *Tyrannosaurus* and a chicken are much more closely related than either of them are to a crocodile. Remember, we paleontologists classify birds as living dinosaurs (avian dinosaurs). Crocodiles and kin, however, are the closest living relatives of birds; they're distant cousins that share a common ancestor, and they are the only living groups of what we call archosaurs ("ruling reptiles"). So, when you think of a chicken and a crocodile, they might not look like each other—far from it—but their evolutionary story is wrapped deep in time.

From a completely different perspective, one thing that birds and crocodilians certainly have in common is that they lay eggs. We know that dinosaurs laid eggs and that some exceptionally rare fossils have even been found of dinosaurs sitting on their eggs, guarding them just as modern dinosaurs (birds) do today, a form of parental care that is also observed in crocodilians.

Living crocodilians typically construct nests near the water's edge and lay their eggs either in holes or mounds, depending on the species of croc. The female will attend to her nest and fiercely guard it against any potential intruders looking to eat the eggs or steal the nest for themselves, so sitting close to or above the nest will aid in protecting the eggs. Remarkably, we have rare evidence of at least one such occurrence in the fossil record.

Despite finding multiple rare dinosaur parents preserved on their nests over the years, the discovery of the first prehistoric croc atop its nest was only revealed in 2015. Considering that some form of parental care is expressed in all living crocs, it could be surmised that such behavior might have one day come to light in a fossil, even though direct evidence would be rare. Well, drum roll, that very fossil was recovered from a coal pit called the "Cecilie pit" within the Geisel Valley of Central Germany, and is about 45 million years old.

The fossil comprised an essentially complete skeleton preserved with at least five ellipsoidal eggs. It was originally collected in 1932 and briefly mentioned in the literature a year later as representing a female croc with eggs, although no behavioral interpretations were provided. Over eighty years later, the fossil was reanalyzed in detail by the paleontologists Alex Hastings, also known as "Dr. Crocogator" (cool name), and Meinolf Hellmund. They identified the specimen as an adult example of a well-studied species called *Diplocynodon darwini*, a type of croc in the alligator family that looked a little bit like a modern caiman.

The skeleton is almost fully intact, including its numerous osteoderms, although there is one odd thing about it: the posture of the body. The body is preserved in an unnatural position, with the head and neck curled around itself, so much so that it looks a bit like a dog chasing its tail. The reasoning behind the unusual posture remains open to debate, although it was suggested that the posture might have been partly due to the female attempting to spread her body over the eggs to ensure they had a more even temperature. Plus, by thoroughly studying other vertebrate fossils, including fish and amphibians collected and carefully mapped alongside this specimen in 1932, it was ruled out that the odd posture was related to the flow of water or rapid burial.

To confirm that the eggs belonged to the adult, a size comparison was made between the adult (51 centimeters long from the tip of the snout but excluding the tail, a measurement known as the snout-vent length) and the eggs (6.6–6.8 centimeters). This comparison revealed that the size ranges observed were consistent with those between two modern crocs (the American alligator and the broad-snouted caiman) and their eggs. Plus, the careful documentation of the fossil site in 1932 revealed that no other croc was found within at least 12 meters of the eggs, providing further support for the claim that this individual was the parent.

Four of the eggs are complete, unbroken, and positioned immediately adjacent to the adult, whereas the fifth egg is incomplete and positioned directly underneath the tail. There are no skeletal remains inside the eggs, suggesting that development was not very advanced. Unfortunately, due to the vertical compression of the fossil, it is impossible to determine what, if any, distance (of sediment) there might have been between the adult and the eggs prior to them becoming fossilized. However, the four eggs bear no signs of having been disturbed and are not damaged, which suggests that they had been buried prior to the mother's death.

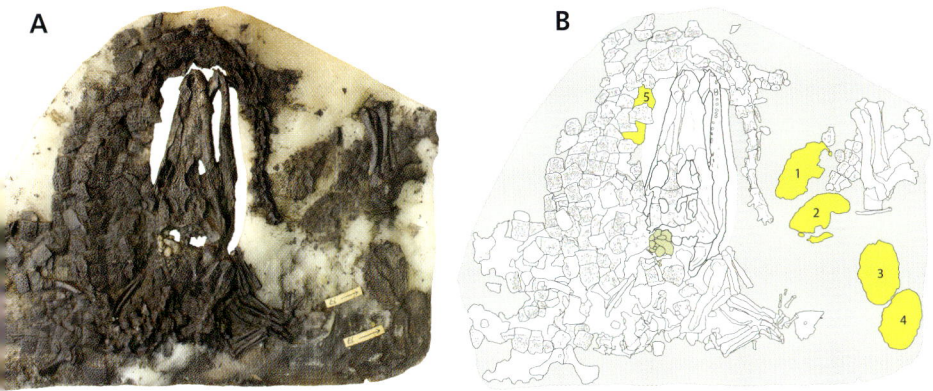

FIGURE 2.2. (A) Photograph and (B) interpretive drawing of the crocodile *Diplocynodon darwini*, guarding her nest of eggs to the very end.

(Courtesy of Alex Hastings)

As Hastings and Hellmund point out, the number of eggs is low for a croc, especially when compared with the typical clutch size of thirty-seven seen in the American alligator, one of the closest living relatives of *Diplocynodon*. They suggested that the eggs may have been laid in a full clutch, and the lack of additional eggs may have been the result of hatching or scavenging, with the broken shells not being fossilized. I guess it is also plausible that this species might simply have laid fewer eggs than is seen in living species. However, the duo hypothesized that this croc mom-to-be may have died atop her nest after oviposition (laying her eggs). But what would have caused her death?

There were no signs of scavenging or any injuries present, and there was no evidence that the death resulted from a chaotic natural event like rapid flooding. Instead, the mother might have died due to direct complications during oviposition, specifically through egg binding (or dystocia), where single or multiple eggs become awkwardly positioned or stuck for a prolonged period inside the mother's oviduct, making it difficult to expel from the body. If correct, it would be the first record in a fossil archosaur. For comparison, dystocia occurs in birds today but is more commonly observed in smaller species and is extremely rare in crocs.

BOB NICHOLLS ART

They also considered that an unexpected drop in the temperature, a cold snap, may have been temporarily too cold for the adult and the young inside the eggs, leading to their demise. Such incidents do occur today. For example, in 2010, a severe drop in temperature in Everglades National Park in Florida resulted in the deaths of at least seventy American crocodiles along with thousands of warm-adapted fish. Studies focused on the fossil plants found in the same area as *Diplocynodon* suggest that the coldest temperature was 16.9–23.0°C, which matches quite well with the temperatures in modern south Florida. Although difficult to confirm, this cold-snap scenario might be a plausible explanation considering the high number of fossil fish found in the same layer as the croc in 1932, which might suggest this was a mass mortality event.

Caring, dedicated parents might not be the first thing that springs to mind when you see a toothy croc, but parental care forms a major part of croc life, from egg laying to hatching and beyond. After the eggs hatch the mom and sometimes the dad will protect the hatchlings for weeks to months; you have probably seen photos or footage of tiny hatchlings hitching a ride on their parent's back. Although we cannot confirm whether *Diplocynodon* similarly cared for its young, while it would seem highly likely, we can say that this unique discovery represents the first and only evidence of an egg-guarding croc in the fossil record. Long live the croc guardian.

FIGURE 2.3. (Opposite). *Casualties of the Cold Snap*

A sudden cold spell has descended upon the wetlands, freezing the water and ground. Sadly, our dutiful mother crocodilian (*Diplocynodon darwini*) and her eggs are also frozen in time.

AMMONITE EGGS

With their often wonderfully intricate spiral shapes, ammonites are some of the most iconic fossils. Thousands of species of these extinct squid relatives have been found globally, which is not really a surprise considering how common these fossils are if you know where to look. To an extent, they are one of those types of fossils that you might think we know everything about simply because we have so many, though that could not be further from the truth. One notable area of ammonite research for which we have little direct evidence is reproduction, which is why this surprising fossil find was so exciting.

To set the scene a little, we must travel back in time, not quite to the Jurassic just yet, but to 12:10 p.m. on Wednesday, September 14, 2011. I was anxiously preparing to deliver a poster presentation at my first-ever academic conference, called the Symposium of Vertebrate Palaeontology and Comparative Anatomy or SVPCA for short. Attempting to lure passersby to look at my poster about a new species of ichthyosaur, one chap suddenly darted over to me and began asking lots of questions and praising me for the research, repeatedly exclaiming, "Good lad, well done." I quickly felt at ease, and my nerves seemed to subside.

Little did I know at that moment in time, but that was Steve Etches, a prolific fossil collector and plumber turned paleontologist. More on that in a moment. Rather befittingly, the SVPCA conference was in Lyme Regis, the historical birthplace of ichthyosaurs, so it was an ideal place for me to speak about them. Etches and I kept in touch, and almost a year to the day, I was preparing to deliver my first professional lecture, this time at SVPCA in Oxford.

Let me tell you, I was incredibly nervous, so much so that I almost decided not to give the talk. I was the last speaker after a long day of lectures, and my talk followed an engaging presentation by Mike Benton, a name familiar to anybody involved with vertebrate paleontology, and who is today a friend and colleague. After overcoming my demons, I delivered the talk, this time discussing a Mary Anning ichthyosaur that I had been studying. Afterward, Etches approached me and said, "That was the best talk of the day! Good lad. Good lad."

My reason for this introduction is to establish the type of person Etches is. He is down to earth, loves his fossils, and is an all-around great guy. His full title is Dr. Steve Etches MBE, and while he trained as a plumber, he started collecting fossils as a kid but continued whenever he had an opportunity. After amassing a truly amazing collection of local Jurassic fossils, primarily from the Kimmeridge Clay Formation in Dorset (approximately 157–145 million years old), his dreams of opening a museum were eventually realized—The Etches Collection Museum of Jurassic Marine Life opened its doors in 2016. I think it is one of the best museums in the UK and one you must visit.

He has discovered, unearthed, and skillfully prepared some amazing fossils, from giant pliosaurs to delicate pterosaurs and fish with preserved soft tissues. He even has a bunch of species named after him. But despite finding so many visually impressive fossils, one of his greatest scientific discoveries—and some might say *the* greatest—is the discovery of the world's first ammonite eggs.

Given Etches's experienced eye for detail and his meticulous collecting ethic, with hopes of finding and collecting as many representative fossils from the Kimmeridge Clay as possible, he was quick to notice anything that seemed unusual. One day, on a routine fossil hunt, he collected and split open a small section of rock that contained two unusual saclike masses of tiny (1–2 millimeters), egg-shaped objects with a golden-bluish nacreous coating. Unable to contain his excitement, on New Year's Eve 1989, at 11:30 p.m. no less, rather than getting drunk seeing the new year in, he got a little merry by studying these finds under a microscope in his workshop.

Puzzled by these strange, three-dimensionally preserved structures, Etches pondered whether they might be fossilized eggs, but from what? Following this idea further, he compared them with modern cuttlefish eggs (a distant relative of ammonites) and noticed they were identical in shape. This led to his thinking that these could be the eggs of an ammonite, given how common ammonites are in the Kimmeridge Clay. Suspected ammonite eggs had been described previously, such as one example from the Triassic of Germany and one from the Jurassic of France, among others. However, more conclusive evidence of ammonite eggs had yet to be found, so this would be a bold claim, especially with just one example. He set to change that. The hunt was on.

Etches sought the help of Michael House, a professor and cephalopod expert at Southampton University, who examined the eggs and excitedly

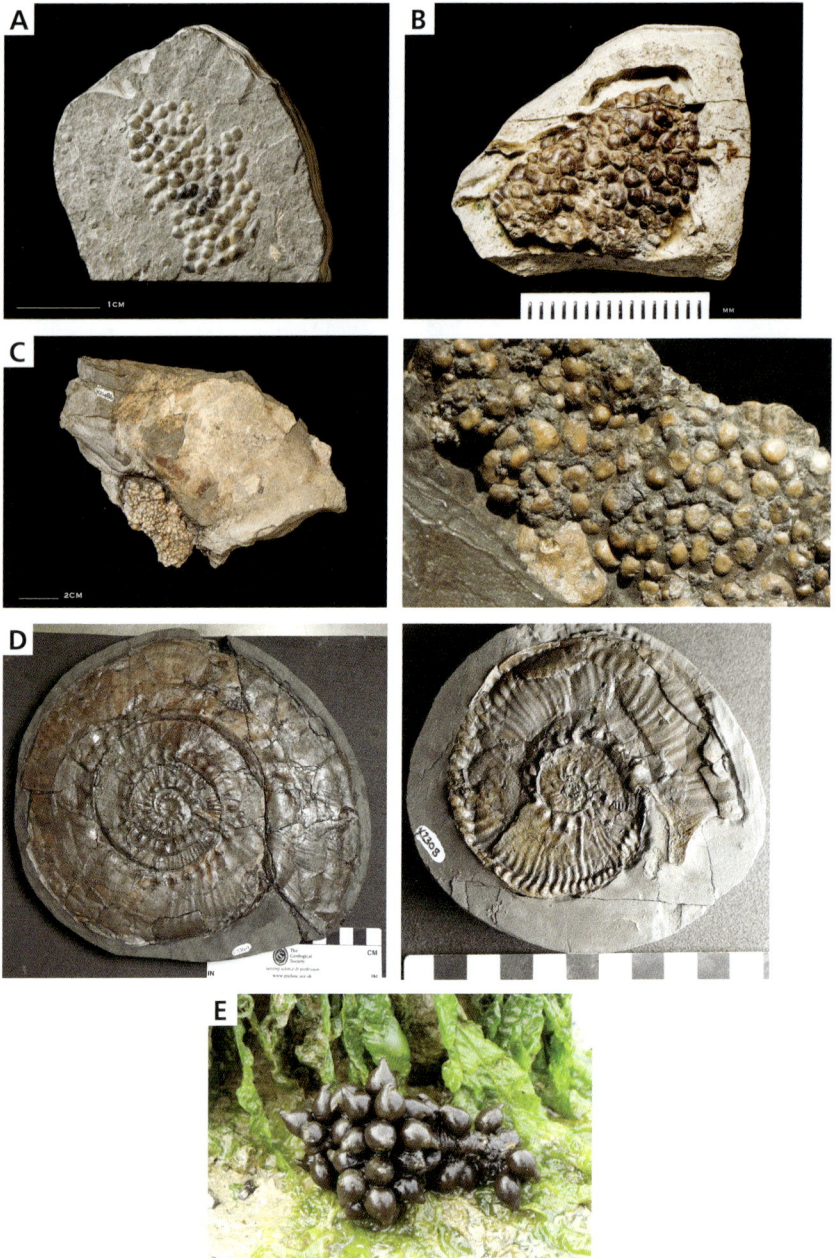

FIGURE 2.4. Ammonite egg sacs, including two (A–B) isolated specimens. (C) An egg cluster associated with the body chamber of an *Aulacostephanus autissiodorensis* ammonite, with a close-up view. (D) Female (macroconch, left) and male (microconch, right) examples of the ammonite *A. autissiodorensis*. (E) Common cuttlefish egg sac washed ashore on the beach.

BOB NICHOLLS ART

agreed that his assertions might be correct. Yet, it was still impossible to say for sure, especially with just that single, unassociated specimen. So, armed with this newfound discovery and having an idea of what to look for, he began searching various locations along the coast. Over several years, he was rewarded with multiple discoveries totaling eight egg clusters, some containing more than thirty eggs. Not only did he and some of his friends find similar examples to the first specimen, but he discovered several sacs of eggs directly associated with ammonites, thus providing more compelling evidence that his ammonite eggs theory could be correct. However, of all these additional finds, the pièce de résistance was the discovery of an ammonite body chamber with a large sac of eggs preserved inside. The eggs were found to be associated with two very common ammonites: *Aulacostephanus* and *Pectinatites*.

Now it was time to write up the findings. To do that, Etches teamed up with two colleagues, Jane Clarke and John Callomon (a world authority on ammonites), who in 2009 described each of the egg samples and explained through a process of elimination that they must be ammonite eggs. Two of the eggs were analyzed using a scanning electron microscope (SEM) at the Geology Department at Portsmouth University, which revealed curious bacteria-like structures inside, although no ammonite embryos could be recognized. It's possible that these two eggs might have been at a stage of initial fertilization, or perhaps they were awaiting fertilization by the male.

Although most people have since agreed that the egg clusters are from ammonites, a few researchers have said that the evidence is not conclusive enough to say 100 percent, although they admitted that it seems most likely. For instance, it might be argued that another cephalopod laid its eggs on or inside an empty ammonite body chamber, similar to how empty mollusk shells are sometimes used today as shelters for animals such as octopods to lay their eggs. There is also evidence of exceptionally preserved, fossilized

FIGURE 2.5. (Overleaf). *A Whole Sac of Babies*

A large female (macroconch) ammonite, *Aulacostephanus autissiodorensis*, deposits her egg sac on the underside of a rock. Her mate, the smaller male (microconch), remains close by and caresses her shell.

gastropod egg capsules from the Jurassic of Russia that were found attached to the empty body chambers of two ammonites, so it is always important to consider every possibility. Indeed, Etches's team considered such scenarios, including whether the parents of the eggs might have been a different mollusk altogether, like a belemnite or gastropod, but based on all the available evidence, especially of those egg-ammonite associations and the structure of the eggs, the team concluded that they must belong to ammonites.

This reminds me of one of my earliest scientific papers, incidentally on ammonites. Working with the geologist Ben Hyde, a friend and former student at the University of Manchester, he and I were writing up our findings of ammonite mouth plates (called aptychi) from near Whitby, Yorkshire, based on discoveries that Hyde had made. Initially, he had found some isolated aptychi embedded in the rock—a bit like Etches's first find—and then he found a couple of others preserved near ammonite shells, which suggested they probably belonged with that species of ammonite. However, when we submitted our research, one reviewer criticized our findings and suggested that we did not have enough evidence that the aptychi belonged to the ammonites because they were not directly associated. So, we headed back to the fossil site to hunt for more, and I found an aptychus *inside* the shell of an ammonite right where it would be expected. Success. Our study was accepted and formally published in 2012. Happy days.

Since his 2009 study, Etches has continued looking for additional eggs and, as you might expect, has found more specimens, including examples from other locations and geologic ages in the Jurassic. His research is ongoing, but the new findings will provide further evidence.

Finding ammonite eggs is quite incredible when you consider that countless ammonites have been found. Unsurprisingly, these little eggs are among Etches's favorite finds. They are also one of my favorite fossils in his museum because they help us to better understand the reproductive behaviors of a group of ancient animals that we already strangely feel like we know, providing us with the best insights yet into the earliest stages in the life cycle of an ammonite.

THAT'S A LOT OF BABIES!

The first mammals, or mammal-like animals, appeared more than 200 million years ago, during the Late Triassic. Because paleontologists often deal with scraps of fossils, the odd tooth here or isolated bone there, it can become quite puzzling when trying to decipher whether one of these early, often shrew-sized animals is actually a mammal or a very close relative. There is a fine line. This is why there is debate among paleontologists on how to define the first *true* mammals. Nevertheless, time spent researching all these early critters is especially valuable as they can provide clues about the origins and evolution of mammals and where to classify them on the mammal and kin family tree.

To go off on a slight tangent, I always like to point out that the famous sail-backed *Dimetrodon*, a firm favorite from the Permian, is not a dinosaur or even a reptile but is closer to the origins of mammals. A distant mammal relative, as it were.

In any event, not to get sidetracked, mammals are part of a much wider group with many animals that were ancestral to them. These are typically referred to as stem mammals or protomammals. They are not mammals by our definition but are related and are part of the overarching group from which mammals came.

One intriguing family of stem mammals is the tritylodontids, which ranged in time from the latest Triassic to the Early Cretaceous. These close mammal relatives lived alongside early mammals and looked a lot like them but did not have all of their key features. However, given that they were related and part of the broader group that would one day lead to humans and the other mammals we see today, any finds that could shed light on their reproduction or parenting would help in our quest to better understand the early evolution and development of mammals.

Enter *Kayentatherium wellesi*, a beagle-sized tritylodont whose fossils come from the 184-million-year-old Early Jurassic rocks of the Kayenta Formation of northeastern Arizona. Formally named and described in 1982, numerous fossils of this creature have been unearthed over the years, but a most curious find was made during fieldwork in June 2000. Collected by the trip leader Tim Rowe, a professor at the Jackson School of Geosciences

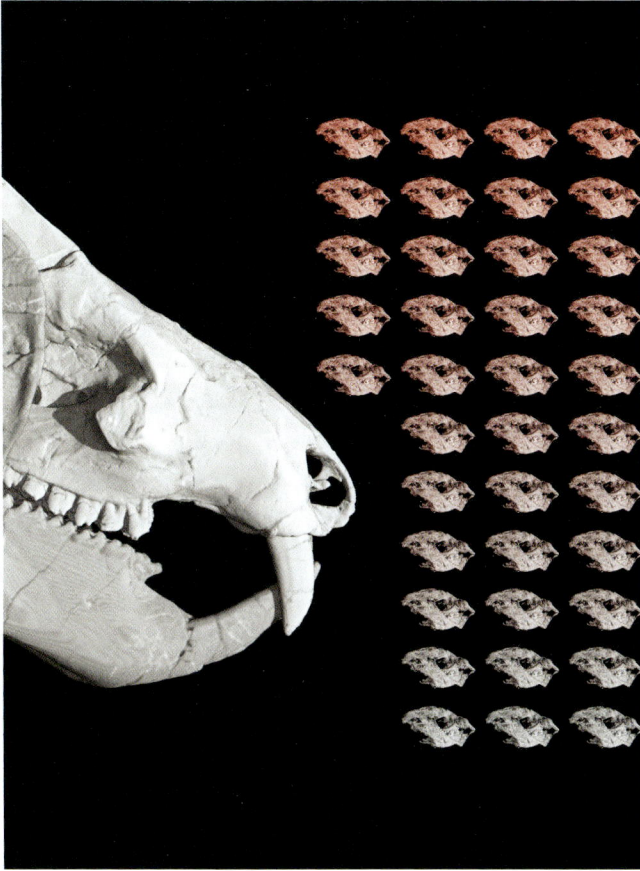

FIGURE 2.6. The adult skull of *Kayentatherium wellesi* compared with the tiny skulls of her clutch of thirty-eight babies found alongside her.

(Courtesy of Eva Hoffman)

at the University of Texas at Austin, the first bones of this *Kayentatherium* were spotted as "surface float" (associated but loose bones eroded out from the rock due to weathering), with more of the skeleton found buried in the ground. When collecting the fossil, Rowe and his team believed they were collecting the remains of a single large adult.

The specimen was one of many to be collected over the years, and it was not until 2009 that a former graduate student and fossil preparator at the

Jackson School, Sebastian Egberts, spotted a tiny speck of tooth enamel. Intrigued by this find, he decided to peer inside the block using a CT scanner. However, it was not until a few years later, with further advances in CT scanning and the processing of enormous amounts of digital data by a graduate student at the university, Eva Hoffman, that enabled a much more detailed look at the specimen and the realization of an extraordinarily rare discovery. Baby *Kayentatherium*. And lots of them.

Excited by the hidden discovery, Hoffman and Rowe studied the specimen and identified an adult *Kayentatherium* preserved with a staggering thirty-eight babies! Their research was published in 2018; this fossilized family represents the only known Early Jurassic mammal precursor with its babies. The remains of the little ones included partial skeletons, jaws, teeth, and ten mostly complete skulls, the latter representing scaled-down, practically identical miniature versions of the adult skull but at a tenth of the size. As it was impossible to say whether they were preserved inside leathery eggs (as in modern monotreme mammals like the platypus) or had recently hatched and were huddled close to the mother's body, Hoffman and Rowe opted to refer to them as perinates, which means an individual from approximately one month before to one month after birth. In terms of how they were buried, the mother and her babies might possibly have been buried inside some type of shallow burrow or den that was dug into the soil.

Modern mammals tend to invest more energy into having fewer offspring and providing some form of parental care, rather than giving birth to lots of babies. Just for your curiosity, the human average is one, but the record number of babies being born at once is nine (nonuplets)! Yet this *Kayentatherium* and her thirty-eight babies are well outside the known range of litter sizes in any living mammal, comprising more than twice the average, showing that it reproduced more in line with what we typically see in reptiles. With such a large brood size, and many mouths to feed, it could

FIGURE 2.7. (Overleaf). *Subterranean Daycare*

An exhausted mother cynodont, *Kayentatherium wellesi*, tries and fails to rest, while her thirty-eight perinates snooze and run wild in the burrow.

be inferred that this stem mammal provided little parental care and that the juveniles may have been self-sufficient at birth, already with functional teeth for munching on vegetation.

Plus, mammals have the biggest brains compared with the rest of the animal kingdom and, indeed, these tritylodontid babies had proportionally smaller brains. This is significant because, as the team nicely put it, this fossil might suggest that a key development in mammal evolution was choosing brain power over brood power. This could be why some of the now-extinct mammal relatives, like tritylodontids, disappeared whereas mammals flourished and became the sole surviving representatives of a once much grander lineage.

Given the Early Jurassic age of this fossil, it provides a unique opportunity for us to study the early reproductive behaviors of stem mammals. Significantly, this is our only snapshot of a close relative near the origins of (true) mammals, so it might help us to understand how our early mammalian ancestors developed an alternative approach to reproduction, helping to change the course of mammalian evolution.

3 | Family and Friends

Crows and magpies use antibird spikes to build their nests, an incredible adaptive response to combat the human-made spikes. Some even tear entire strips from buildings and use the spikes like humans intended—to ward off other birds. Smart dinosaurs.

THE MOTHER IN THE TREE

You are a synapsid, just like a dog, a cat, or a dolphin. Don't worry—it's not an insult, just a fact when classifying animals using scientific terminology. All mammals, including humans, belong to the synapsids, one of two major groups. This also includes the numerous groups of mammal-like animals, and protomammals, which came before true mammals, including the likes of *Dimetrodon*, which we briefly touched upon in the last chapter. The other major group, or sister group, to the synapsids is known as the sauropsids; it includes all reptiles, birds, and many of their extinct ancestors.

Synapsids and sauropsids are both amniotes, tetrapod vertebrates that evolved the amniotic egg (or "land egg"), which freed them from the aquatic realms. The oldest amniotes appeared a little over 300 million years ago during the Carboniferous, when they began to rapidly diversify. This was a pivotal time for the evolution of the earliest synapsids, the ancestors of mammals, which allows us to ponder what types of behaviors improved their chances to succeed and pave the way for their success.

Notably, parental care is a common behavior seen among mammals today. Think about humans caring for their children at all costs, dogs fiercely protecting their newborn puppies, or elephants teaching their young to find food or fend off predators. Although it can be costly to the parent, investing in parental care is investing in the future and enhances the chance of the offspring's survival to adulthood. But where does this behavioral instinct and strategy to care for offspring come from, and when did it evolve in the ancestors of mammals? As might be expected, finding evidence for such behavior is inherently rare, but a recent discovery of an exciting synapsid fossil has helped to shed some light on such prehistoric parenting.

The newly discovered fossil is between 309 and 306 million years old and belongs to an early group of synapsids called varanopids. Lizard-like in appearance, they kind of resemble living monitor lizards (whose scientific name is *Varanus*, hence the name varanopids) and include species that are among the oldest known tree-climbing (arboreal) vertebrates. Their fossils have been found in North America, Russia, Europe, and South Africa, and this specimen was collected from Cape Breton Island, near the town of Sydney in Nova Scotia, Canada.

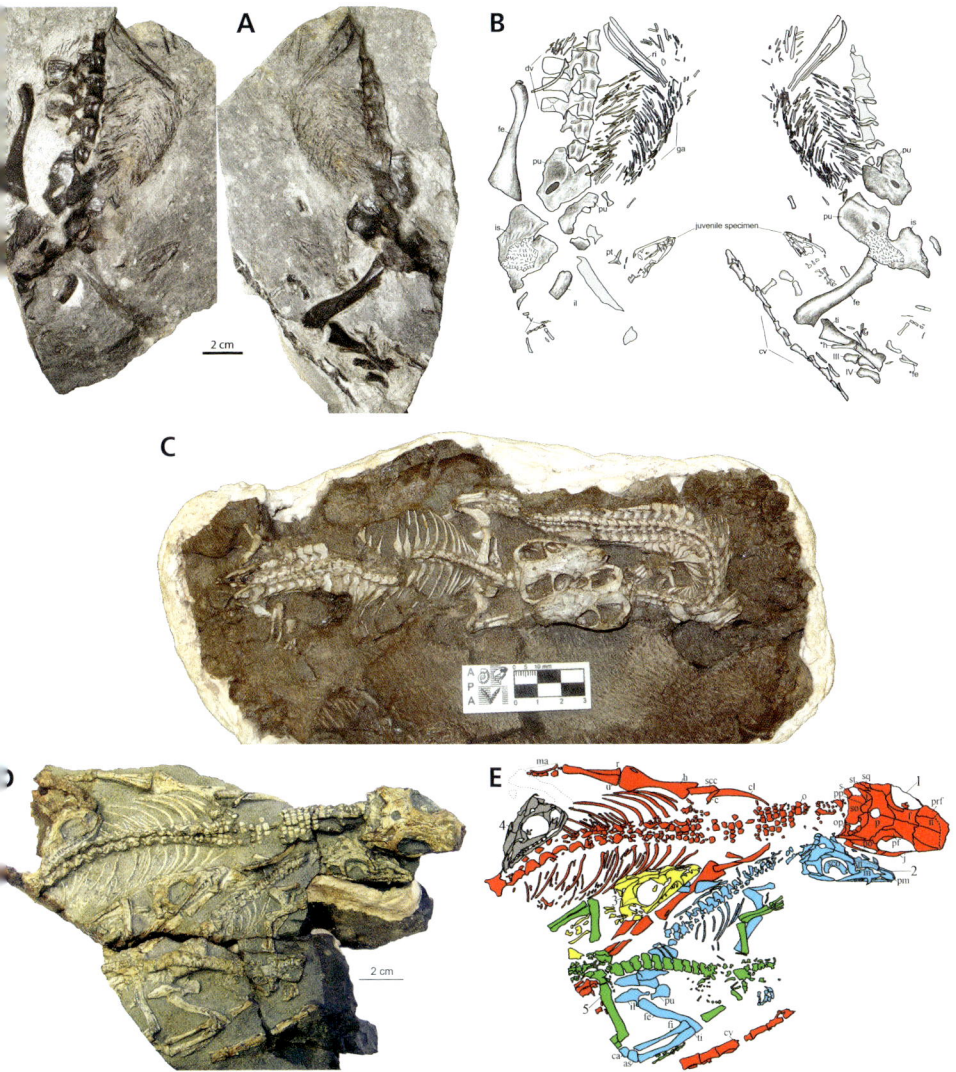

FIGURE 3.1. (A) Photograph and (B) interpretive illustration of the *Dendromaia unamakiensis* adult and juvenile. (C) A pair of late juvenile *Thrinaxodon* with their heads resting together. (D) Photograph and (E) interpretive illustration of the *Heleosaurus* family huddled together. The red individual is the adult, which is clearly larger than the four juveniles.

([A–B] Courtesy of Hillary Maddin; [C] image from S. C. Jasinoski and F. Abdala, "Aggregations and Parental Care in the Early Triassic Basal Cynodonts *Galesaurus planiceps* and *Thrinaxodon liorhinus*," *PeerJ* 5 (2017): e2875; [D–E] courtesy of Jennifer Botha)

The fossil contained the remains of an adult and juvenile varanopid found nestled inside the stump of a giant fossilized lycopsid tree (an extinct relative of modern clubmosses), from a time when immense coal forests spread far and wide. Announced to the public in December 2019, just in time for Christmas, the pair were also identified as a new genus and species, called *Dendromaia unamakiensis*, the genus name after the Greek *dendron* meaning "tree" and *maia* meaning "caring mother," in light of the association.

The identification of these specimens as a new species was exciting, even more so because they represent the only varanopids known from Nova Scotia and the earliest record of the varanopid family, predating the previous record holder by 8–10 million years. As important as that is for varanopid research, it was the adult-juvenile association that really sparked the interest of the research team, led by Hillary Maddin of Carleton University in Ottawa, Canada, along with Arjan Mann, and the citizen scientist Brian Hebert, who discovered the specimen in 2017.

Evidence for parental care in fossil vertebrates is generally limited to the preservation of directly associated individuals of the same species but of different ages. As preserved in this fossil, the adult and juvenile are represented by two articulated partial skeletons. The larger individual is preserved as part and counterpart (the matching halves of the same fossil) and comprises the back side of the skeleton, whereas the smaller individual includes a tiny, 3-cm-long skull and associated partial skeleton. The vertebrae of the latter are 25 percent the size of the same vertebrae in the large individual; in other words, the adult is four times larger than the juvenile. Excluding the long tail, the adult had a body length of around 20 centimeters.

It is obvious that the pair are members of the same species (conspecifics), and presumably mother and her youngling, not least because of their close association but because of their overlapping anatomy being indistinguishable from one another. Looking more closely at their position, the smaller

FIGURE 3.2. (Opposite). *Their Watery Grave*

Caught in a powerful Carboniferous storm, a mother lizard-like varanopid, *Dendromaia unamakiensis*, and her youngling take shelter inside a lycopsid stump. However, the rain will continue to fall, and their shelter will become their grave.

individual is neatly tucked beneath the left femur of the larger individual, where it is surrounded by its long tail, suggesting the adult was protecting the juvenile. This position is reminiscent of what would be found among denning animals, just like today, indicating that the tree stump in which they were found was probably also their den. As both individuals are well articulated and such delicate bones as the belly ribs (gastralia) are preserved, this suggests that they were rapidly buried with little or no transport, meaning that they pretty much died where they lay, presumably resulting from a storm that flooded the swamp-like forest and their stumpy home.

Based on the association, position, and preservation of the pair, Maddin and her team proposed that two forms of parental care might have been in place: prolonged offspring attendance, where the adult cares for the young for an extended period of time following birth, and offspring concealment, where the young are kept hidden, in this case in the tree stump den. The level of investment and protection for the young can be exhausting and costly to the parent, which will go to great lengths to ensure that the young are well nourished. Such parental care significantly increases the survival rate of the offspring, with the parent providing food and protection during early development.

Further supporting evidence for parental care comes from additional synapsid fossils, including multiple aggregations of the mammal-like cynodonts, *Thrinaxodon* and *Galesaurus*, found in Early Triassic rocks of the Karoo Basin of South Africa. The only other record of parental care in a synapsid comes from yet another varanopid, called *Heleosaurus*, which comprises a family frozen in time from the Middle Permian of South Africa. Supporting the interpretations above, in this *Heleosaurus* specimen, four articulated juveniles are lying in a lifelike position alongside and nestled up to the articulated body of a parent, whose tail also encircles the juveniles, all of which were probably buried inside some form of den or shelter. The four juveniles are identical in size and represent siblings that are about 50–60 percent of the adult's size, which suggests parental care in *Heleosaurus* was rather extensive and continued beyond the postnatal stage.

The adult and juvenile association observed in this fossil is the oldest evidence of parental care in a synapsid, predating the previous record holder by 40 million years. Sitting far back on the evolutionary tree (or, in this case, in the stump of it), deeply rooted near the early diversification of synapsids, this fossil provides the earliest starting point to better understand the origins of parental care in our distant mammalian forebears. This seemingly critical form of behavior would go on to play a significant role in mammal evolution. Looking at it this way, it is incredible to think that parenting has been part of our ancestry for more than 300 million years.

WHAT YOU (AND I) DIDN'T SEE AT EGG MOUNTAIN

In 2009, I was part of a team helping to dig up a dinosaur bonebed containing multiple hadrosaurs (*Maiasaura*) and a tyrannosaur (*Gorgosaurus*) just a few miles away from the tiny city of Choteau, Montana. If you're a *Jurassic Park* buff, you may recognize the location. In the movie, Hammond exclaims, "I've got a jet standing by at Choteau," after popping open a bottle of champagne to celebrate Alan and Ellie agreeing to come with him to his park. In return, Hammond agrees to "fully funding" their dig for a further three years. Let me tell you, being offered three years of funding like this is harder to believe than cloned dinosaurs! As it turned out, working in the Montana badlands, I was about a fifteen-mile drive from a world-famous dinosaur site known for the discovery of dinosaur eggs and babies: Egg Mountain.

Unfortunately, while I didn't get a chance to visit the Egg Mountain area, I did visit The Montana Dinosaur Center to discover more and meet a paleontologist, Dave Trexler, whose mother, Marion Brandvold, found the first remains of baby duck-billed dinosaurs in 1978. Brandvold had a lifelong fascination for all things fossils after finding her first dinosaur bone in 1917, at the age of five, and in 1937 she opened the Trex Agate Shop in Bynum, just north of Choteau.

Her discovery in 1978 was groundbreaking. Showing the remains and the site to the paleontologists Jack Horner and Bob Makela led to the identification of a dinosaur (*Maiasaura*) nesting ground, which—for the first time—demonstrated that at least some species lived together in colonies and cared for their young. Additional nests were even found buried in rock layers one above the other, showing that *Maiasaura*, or the "good mother lizard," used the same nesting grounds year after year. Further fieldwork in this fossiliferous area known as the Willow Creek Anticline led to the discovery of so many fossils, including hundreds of eggs and many hatchlings, that one part of the area was dubbed "Egg Mountain." The site is officially within the Beatrice R. Taylor Paleontology Research Site, named in 2004.

The discovery provided the first convincing evidence that dinosaurs exhibited complex behaviors, in this case parental care. As you might expect, combining caring parents and baby dinosaurs together, the research

FIGURE 3.3. (A) The "good mother lizard," *Maiasaura*, feeds her hungry hatchlings, as displayed at the Wyoming Dinosaur Center. (B) Photograph and (C) interpretive illustration of some adult and subadult *Filikomys primaevus*, a rodent-like multituberculate mammal found at Egg Mountain.

([A] Photograph by the author; [B] courtesy of Luke Weaver; [C] courtesy of Misaki Ouchida)

was covered widely in the news and is to this day often included in dinosaur books and TV documentaries. Over the years, most of the research undertaken at this site has focused on the dinosaur nesting grounds, but a new study in 2020 revealed that another type of animal lived alongside the dinosaurs in their very own nesting grounds: mammals.

Since the original findings in the 1970s, a great deal of fieldwork has been undertaken around the Egg Mountain locality, which is Early Cretaceous in age and part of the Two Medicine Formation, being about 75.5 million years old. Not only has this led to the discovery of many more dinosaur bones and eggs but also the remains of lizards and mammals and even insect trace fossils including burrows, cocoons, and pupation chambers. Such discoveries

paint a rather vivid picture of a lively dinosaur nesting community with a variety of animals living in and around the nesting grounds. However, although mammal remains had been found in the area before, a significant collection was unearthed during the 2010–2016 field seasons, which provided the first insights into the social behavior of a Mesozoic mammal.

The mammal in question is called *Filikomys primaevus*, a rodent-like species belonging to a now-extinct group known as multituberculates. This mouse-sized mammal was formally announced as part of the 2020 study, which was led by Luke Weaver, a then-graduate student in biology at the University of Washington. These specimens comprise semi-articulated skulls and skeletons, including the first associated multituberculate skeleton from the Mesozoic of North America. The study focused on the occurrence of multiple individuals that were found in tightly packed, monospecific aggregations that suggest some form of sociality.

A minimum of twenty-two *Filikomys* individuals were found and recorded in distinct groups within the same general area at Egg Mountain. Some of the groups contained multiple individuals of mixed ages, including subadults and adults. Finding groups of *Filikomys* buried together suggests that they were living belowground, presumably in some type of burrow. Being concealed inside a burrow would be the best explanation for why the remains were so well preserved, although no burrow system has yet to be found. With that said, based on the geology of the site, the researchers inferred that any sort of burrow would be a challenge to identify or even preserve.

Additional support for the burrowing inference comes from the anatomy of *Filikomys*, which shares several features with living burrowers, such as having powerful shoulders and elbows, and is most analogous to the least chipmunk. It is also worth noting that thirteen of the individuals were preserved in four groups found in close proximity, which suggests that they might have been living in colonies, reminiscent of various burrowing rodents and rabbits today.

FIGURE 3.4. (Overleaf). *Living in the Egg Mountain Shadows*

Beneath the feet of nesting giant dinosaurs, *Maiasaura peeblesorum*, a family of mini multituberculate mammals, *Filikomys primaevus*, thrive in their burrows and among the eggs, dung, flakes of skin, and plant detritus that fall from above.

Further to this scenario, there is no evidence at the site to suggest the skeletons were transported by a river and subsequently buried or any evidence of bite marks or acid etching that might suggest they were individual meat caches collected and stored by a predator. Rather, the excellent preservation of the fine and delicate bones, combined with aggregations of the same species and of different ages, suggests that *Filikomys* nested together in groups or families. In fact, one of the main aggregations comprised three adults and two subadults, suggesting that this was more likely to be gregarious social behavior than parental care.

Interestingly, several of the *Filikomys* groups were located at different geological levels at the site, representing different time intervals. This indicates that generations of *Filikomys* engaged in group nesting and burrowing behavior, utilizing the same areas year after year, a bit like the *Maiasaura*. Emphasizing the importance of these *Filikomys* fossils, they represent the most complete and well-preserved Mesozoic mammal remains from North America, the first documented occurrence of adults and subadults preserved together, and the oldest evidence yet of a burrowing multituberculate. Quite the discovery.

You can easily imagine these mini mammals living around dinosaur nesting grounds, quickly darting inside their burrows at any sign of danger, living their lives underneath the feet of dinosaurs. I like the fact that *Filikomys* means "neighborly mouse," which is quite apt. It is also quite profound to think that one day another type of social mammal—humans—would come along and discover one of the most important dinosaur sites in North America famed for the evidence of dinosaur social lives, only for it to later reveal the first evidence of group sociality in a Mesozoic mammal.

THE WATERY GRAVEYARD

About 76 million years ago, a huge herd of ceratopsian dinosaurs was about to have a *really* bad couple of days. Droves of ceratopsians for as far as the eye could see were hastily moving along the coastal plain, frantically barging each other out of the way, striking into trees and struggling to clamber across slippery slopes and wade through rapidly rising rivers. The little ones were being pushed aside and trampled, squished into the sloppy mud, and separated from their close-knit family groups. Bellows and groans called out. Panic was in the air. It was chaos.

But what caused this surge of all-out alarm? A predator, perhaps? No, it was something far more ominous. A tropical storm, a hurricane, hit the coastline, and surging water levels triggered severe flooding along the coastal plain. The home these dinosaurs once knew was rapidly disappearing beneath the rising waters, where soon the entire landscape would be radically transformed and become unrecognizable. With no high ground nearby to flee to for refuge, these ceratopsians were running for their lives and running out of time.

Sorry to toy with your emotions there and paint such a vivid image of a fateful day in the Late Cretaceous, but this is based on a significant fossil site that was first discovered, at least in part, in the late 1950s and 1960s and studied extensively in 1997 by paleontologists from the Royal Tyrrell Museum.

Situated in the Hilda area of southern Alberta, Canada, there is a magnificent series of at least fourteen bonebeds containing the rhino-sized horned ceratopsian *Centrosaurus apertus*. Although dozens of centrosaur bonebeds have been documented in Alberta, these are special because they are identical and are preserved in the same age layers of mudstone, together forming what is identified as a mega-bonebed, in this case the "Hilda mega-bonebed," an extensive site that covers an estimated area of at least 2.3 square kilometers based on the distribution of the fourteen bonebeds. That is an area about the size of three hundred soccer fields.

Reviewing the sheer volume of bones, the cumulative minimum number of individuals estimated in this mass graveyard is in the very low thousands, representing a truly enormous dinosaur herd that perished together during a disastrous flooding event. Combined as one, this site is one of the largest dinosaur (if not the largest) bonebeds in the world and records a genuine scene of catastrophe.

If we momentarily forget about the wave of devastation, these dinosaur-rich rocks confirm the herding nature of this species, providing direct evidence of sociality in an enormous group. Presumably, with comparisons to modern-day mammal herds, these ceratopsians must have lived in distinct family groups within the wider herd. Further evidence for such interpretation comes directly from the bonebed, which has yielded a mixed range of individuals, including juveniles and adults. Data from the site also suggests that these centrosaurs exhibited some form of east-west seasonal migration behaviors, which paints quite a dramatic scene of centrosaurs

FIGURE 3.5. The author with science communicator Jimmy Waldron (of the Dinosaurs Will Always Be Awesome traveling museum, at left) provides scale for an adult skeleton of *Centrosaurus* at the Raymond M. Alf Museum of Paleontology.

migrating across the landscape. The site represents one of the first discoveries to provide substantial evidence of sociable behavior in the dinosaur fossil record.

The bonebed is a mishmash of jumbled, disarticulated skeletal remains with rarely articulated or associated individuals. If it was not enough for these dinosaurs to suffer such a terrible death, when conditions finally turned more favorable and the floodwaters began to recede, those few members that somehow managed to survive were surrounded by a sea of ceratopsian carcasses claimed by the storm. The sad fact is, as studies and interpretations of the site have revealed, the bodies were not buried directly as the result of this storm but were tossed around and eventually submerged by subsequent flooding events that happened months to years later. This area seemed prone to monsoonal-type seasonal storms that created perilous conditions each year. Tragically, we see these types of disastrous flooding events devastate low-lying coastal areas today, such as Bangladesh, often claiming the lives of humans and many other animals.

Some evidence for the exposure of the carcasses to the elements and a lack of rapid burial is contained in the bones themselves, with many showing evidence of weathering, trampling, and even scavenging, although that is rare. Many bones are broken and fragmentary. Of course, these ceratopsians were not the only animals living in this area at the time of their demise. Aquatic animals including fish, turtles, and crocs have also been found at the site, as has an abundance of plant remains. Many other species, especially much smaller ones, would have been able to scamper to some form of safety; perhaps mini mammals and lizards climbed trees and hung on for dear life, whereas birds simply flew away. Isolated, shed teeth of large and small theropods were also recorded, as were a few scratch and tooth marks found on some of the *Centrosaurus* bones, presumably left by hungry theropods that came for a free centrosaur smorgasbord.

FIGURE 3.6. (Opposite). *The Day of the Deluge*

Hundreds of panicked ceratopsian dinosaurs, *Centrosaurus apertus*, both young and old, are engulfed by a catastrophic flash flood. Few will survive.

For this enormous herd of ceratopsians, opting to spend at least part of their time along the coastal lowland would ultimately be their downfall. Given the low-level topography of the wetland-like landscape, such a location is highly vulnerable to the impact of flooding due to extreme weather, thus rendering these animals at a serious risk of eradication. During a severe tropical storm, not only would there be an abnormal surge in sea level, but the heavy rains would cause rivers and lakes to burst their banks and quickly overflow with nowhere for the water to go. The area would have become swamp-like, with plants and tree trunks submerged. Unfortunately for these sizable dinosaurs, they could not outrun the floods and could not swim for great lengths or head to higher ground nearby. There was nowhere to escape, and a watery death loomed large.

Despite the tragic tale this mega-bonebed reveals, it shows that these ceratopsians lived at least part of their lives in gigantic herds not too dissimilar to some large mammal herds today. Perhaps more significantly, it provides some of the most compelling evidence for sociality and parental care in horned dinosaurs.

ARMOR AND OUT

If you ever find yourself traveling down the fantastically named Gooloo-gong Road between the small village of Gooloogong and the small country town of Canowindra in New South Wales (NSW), Australia, then know that you are in the vicinity of a discovery hailed as one of the most significant fossil finds in Australia.

As any paleontologist will tell you, and as you may already know, many great fossil finds are made by chance, often by people simply in the right place at the right time. A hiker walking in the hilly countryside, your friend's uncle digging up his garden, or somebody taking a whizz behind some boulders and finding a fossil (it happens, believe me). This chance discovery, however, was made between 1955 and 1956 during roadworks, following the local council's decision to remove a dangerous rocky corner on Gooloogong Road just six miles west of Canowindra.

A heavy-duty bulldozer was used to clear the way and smash through massive sandstone slabs, some weighing more than two tons, and push them over the side and down the embankment where they would be buried. One of the slabs measuring 1 × 2 meters seemingly caught the eye of a bulldozer operator, Charlie Stevens, who was intrigued by some odd markings in the block and had that piece placed alongside the road near a fence. Nothing more was to be made of it until months later when the local beekeeper Bill Simpson and his foster son Bob Scott happened to see this block at the side of the road and sat on it, waiting to be picked up. Recognizing that some sort of fossils were preserved in the block, Bill wrote a letter addressed to the Australian Museum in Sydney, about two hundred miles away, with a line that read " . . . with what looks like a fossil reptile on it."

Intrigued by the report and coincidentally planning to be near the area, the then-museum paleontologist Harold Fletcher and Ted Rayner from the NSW Geological Survey visited the area to inspect the slab. Although the roadworks had long been completed and the original fossil layer covered over, they immediately realized the potential scientific value of the find. They even cancelled the rest of their trip to focus on the fossil and arrange for it to be safely moved to the museum in Sydney for further analysis and eventual display.

See, this (re)discovery was not of a fossil reptile but a slab containing more than a hundred armor-plated fish, called placoderms, preserved together in a mass death assemblage. Placoderms were among the first fish with jaws, evolving more than 400 million years ago; perhaps the most famous placoderm and a name you might recognize is the large-bodied, big-mouthed *Dunkleosteus*, which was one of the first large apex predatory fish to evolve.

Preserved in this giant block, which came to be known as the "Canowindra slab," were a total of 114 fish represented by four different species. The block was heavily dominated by two different kinds, called *Remigolepis* and *Bothriolepis*, members of a successful group of placoderms known as antiarchs, and both were well-known from fossils in the northern hemisphere. The remaining fish included two juveniles of a rare placoderm called *Groenlandaspis*, found previously in East Greenland in the 1930s, and there was also an entirely new type found in the center of the block. This was not a placoderm but instead a single, complete, and beautifully preserved lobe-finned fish (a sarcopterygian, a member of the same wider group that includes coelacanths and lungfishes). Named in 1973 as *Canowindra grossi*, this specimen was the first complete sarcopterygian found in the southern hemisphere.

Finding so many individuals of several species preserved in a single block, each with the same level of preservation and in the same layer, captures a tragic moment that led to the mass death of a community of ancient fish some 363 million years ago, during the Late Devonian. But this slab had only scratched the surface of what was to come.

Following the fossil's move to Sydney, the actual site was virtually forgotten until Alex Ritchie was appointed the new paleontologist at the Australian Museum in 1968. Ritchie had spent many years studying fossil fish, even completing his PhD on fossil fish from Scotland, and he was fascinated with the Canowindra slab, so much so that he contemplated the idea of relocating the original site. Building on his excitement, he visited Canowindra at least six times between 1973 and 1990 to attempt to relocate the exact layer where this slab had come from but without success. Help and

FIGURE 3.7. (Opposite). (A) The one that started it all: The Canowindra slab contains 114 fish, including four different species of placoderms. (B) Excavation of the site in July 1993. (C) Some of the more than sixty tons of slabs collected during the dig. (D) One of the example slabs collected in 1993, including densely packed specimens of *Remigolepis* and *Bothriolepis*.

(Courtesy of Mackenzie Enchelmaier at the Australian Age of Dinosaurs Museum of Natural History and the Age of Fishes Museum)

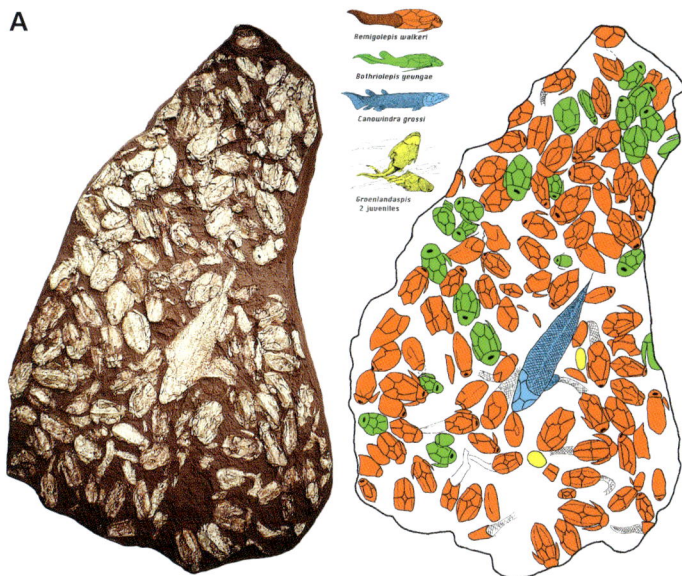

A

Remigolepis walkeri

Bothriolepis geungae

Canowindra grossi

Groenlandaspis
2 juveniles

B

C

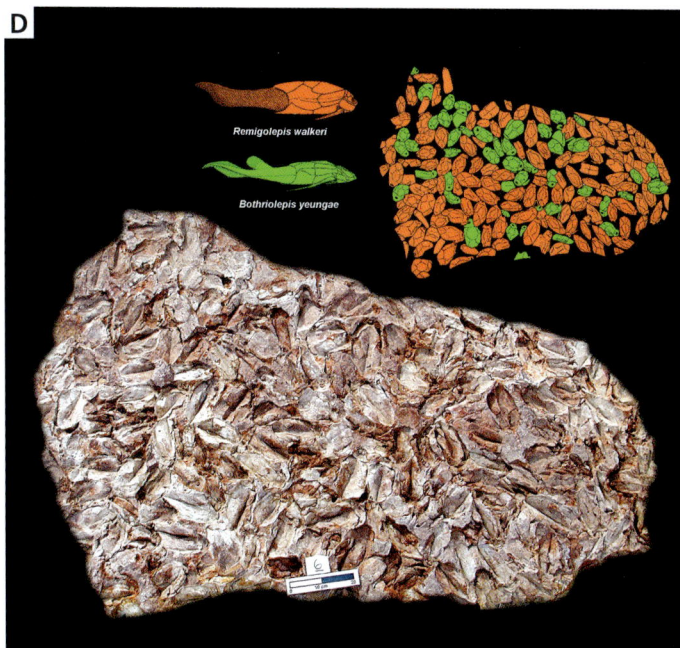

D

Remigolepis walkeri

Bothriolepis yeungae

BOB NICHOLLS ART

a touch of fate came from an unlikely source, a Sydney dentist called Bruce Burns, who had seen the fossil slab on display at the Australian Museum in 1956 and had recently bought property in Canowindra. Long story short, Ritchie was eventually given access to heavy-duty equipment free of charge by local council members to try to relocate the site.

A trial one-day dig was undertaken in January 1993 just to see if there was even the slightest chance that the original fish layer could be located. The result? It took less than three hours before the layer was found. The dig far exceeded any expectations and was so successful that every slab the excavator turned over was crammed with complete fish fossils. Ecstatic with the results, Ritchie formed a team and returned to the site in July 1993 to undertake a major excavation with the help of the council and the Canowindra community.

Incredibly, the team unearthed more than 60 tons of fossil slabs containing almost four thousand individual fish preserved in the same layer of rock! Not only that, but nearly all of the specimens were complete, piled next to one another, and several additional species were identified. Throughout the thousands of fish in this single community, the dominant species were still *Bothriolepis* and *Remigolepis*, making up a staggering 97 percent of all known fish, and about fifty to sixty additional specimens of *Groenlandaspis* were found, both juveniles and adults. Plus, along with the very rare *Canowindra*, which is still only known by one specimen, at least five types of air-breathing lobe-finned fishes were also discovered, including species never seen before, such as *Mandageria fairfaxi*, the largest species up to 1.6 meters long, *Cabonnichthys burnsi* named in honor of the local Cabonne Council and Bruce Burns, *Gooloogongia loomesi* (I love that name!), and *Soederberghia simpsoni*, a lungfish named in honor of Bill Simpson.

This amazingly rich treasure trove became known as the Canowindra fossil site, which provided a direct snapshot 360 million years into the deep

FIGURE 3.8. (Opposite). *Victims of the Parched Pools*

As the water in the oxbow lake evaporated and drained away, the fish became increasingly concentrated in the shrinking mud pools. Eventually, no water remained, and the multitudes perished, including *Remigolepis, Bothriolepis, Groenlandaspis,* and *Canowindra.*

past. The cause of this catastrophic mass death event is still under investigation, although it appears to have been the result of a drying oxbow lake (or billabong in Australia), where a large area of water is separated from the main river as it changes its course, thus trapping the occupants in a freestanding body of water and therefore at the mercy of the weather. In this case, the fish community were exposed to a severe drought and ultimately perished en masse before they were covered by fine sediment and buried.

You might be wondering, "Where are you going to put sixty tons of fossil fish?" Legit question, right? Following on from the grand success of the 1993 dig, Alex proposed that a museum should be built in Canowindra to showcase the fantastic fishy finds.

Fast forward to February 2000 and to the grand opening of the Age of Fishes Museum, filled with specimens collected from the site. As a nice touch, the original slab, which was previously on display at the Australian Museum for fifty years, was returned to Canowindra in 2006 for permanent display in the new museum. To this day, the museum continues to share the fossil wonders of Canowindra with the public and has plans to build an on-site attraction at the original location to continue excavations, with hopes of maybe discovering some inherently rare early tetrapods. After all, who knows what might one day turn up, considering that many thousands more specimens must still be buried at this remarkable mass fish grave.

4 | Moving Along

The walking catfish can leave the watery world behind for up to eighteen hours and can comfortably "walk" on land for 1.2 km. Records exist of groups emerging from drains to hunt worms in parking lots. Some armored species walk so strangely that a new term was coined for this behavior: reffling. It certainly puts a spin on the old saying "a fish out of water."

SWIMMING DINOSAURS

In April 2022 the trailer dropped for the highly anticipated TV series *Prehistoric Planet*. One of the major standouts from the trailer was a swimming *Tyrannosaurus rex*, an image that was also plastered on billboards in Los Angeles. Social media was filled with comments about the swimming *T. rex*, which was great marketing for the series as it was a case of selecting something that is well known, in this case the most iconic dinosaur of all, and showing it doing something totally out of the ordinary, for which everyone loses their minds—although the out of the ordinary was indeed something very ordinary: swimming.

Why not a swimming *T. rex*? Ostriches, kangaroos, and sloths are capable swimmers, but you probably would not think it if you looked at them. Even bigger beasties like moose and elephants swim. Plus, we know that many dinosaurs had a complex system of air sacs in their bodies, like birds do, making them much lighter and providing buoyancy for swimming.

Of course, we know that many dinosaurs would have spent time in the water, be it foraging for food, crossing a river, or hopping from island to island, but no dinosaurs lived exclusively in the watery world, contrary to what a certain book has claimed. Even today, the most aquatically adapted dinosaur of all, the penguin, does not live exclusively in the water. Just in case you were wondering, but I doubt that you were, ichthyosaurs, plesiosaurs, and mosasaurs did spend their lives in the water, but these creatures are not dinosaurs. If you need a speedy refresher on this, then please check out my book *Dinosaurs: 10 Things You Should Know* because this is absolutely one thing you should know.

No doubt most dinosaurs could swim, at least a little, and some of them were probably much more adapted to life in the water than others. For instance, sail-backed *Spinosaurus* appears to have had some adaptations that suggest it spent at least part of its life in the water. However, we are not about to dive into the aquatic *Spinosaurus* debate (maybe some other time). Rather, we will stick with more solid evidence in the form of fossilized swimming tracks. As might be expected, swimming dinosaur tracks can be tough to interpret and are generally pretty rare, more so because dinosaurs did not spend all of their time in the water. So, it would not necessarily have been a common, daily behavior that could be so easily and frequently captured in the fossil record. But also, how could swimming be preserved as a track anyway?

Bringing it close to home, at least for me, a key study that focused on dinosaur swim tracks was based on a specimen from Middle Jurassic rocks from along the Yorkshire Coast in England. In chapter 1 in the section "Mega Millipede Mating," you might remember my comments about the paleontologists Martin Whyte and Mike Romano, who dedicated much of their careers to studying dinosaur footprints from the Yorkshire Coast. This led to the recognition of this area as one of the best dinosaur track sites in the world.

In 2001, the pair described a trackway that consisted of at least six distinct footprints arranged in pairs that were defined by three subparallel ridges with a depression at the back end. They determined that the three ridges were deep scratch marks produced by a swimming dinosaur propelling itself forward in the water, scraping its toes through the underlying substrate and penetrating the layers below, a bit like a rake being pulled through sand. Based on the three-toed type of track, along with the identification of many tridactyl (three-toed) tracks assigned to theropods and ornithopods from the same area, this swim track was probably made by a small theropod taking a dip. They called this new type of rake-like dinosaur track *Characichnos tridactylus*. Many examples have since been found in the area, and I even found and identified one during my early twenties.

Since the original description, several reports of *Characichnos*-type swim tracks have been published, including at major dinosaur track sites of different ages and from around the world, like the St. George Dinosaur Discovery Site at Johnson Farm in southwestern Utah. At this Early Jurassic site, which is about 198 million years old, a huge collection of more than 3,200 well-preserved dinosaur swim tracks were identified, including those created by dinosaurs swimming down-current, up-current, and cross-current. The site is the largest and best-preserved of any known dinosaur swimming track site on record. The preservation is so good that even sharp claw impressions that penetrated the muddy lakebed are present as well as, in some places, details of the footpads and skin.

All of the tracks were produced by small—and some medium-sized—theropods that were swimming in the shallow waters of a large, ancient freshwater lake called Lake Dixie, which was about sixty miles across. The great abundance of swim tracks strongly suggests that large groups of small

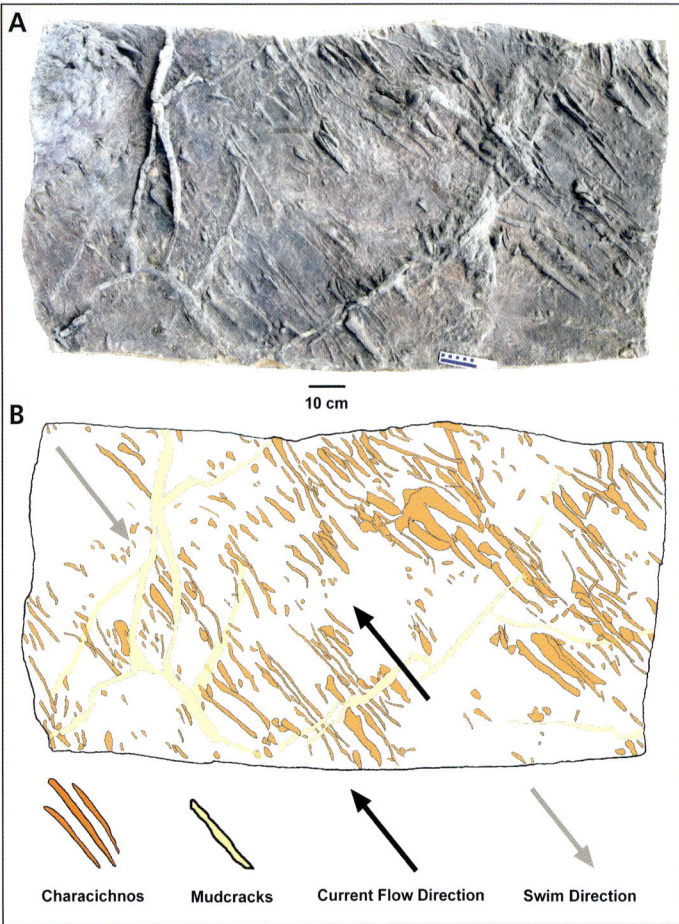

FIGURE 4.1. (A) Photograph and (B) interpretive drawing of a high-density, three-toed swim track assemblage from the Early Jurassic St. George Dinosaur Discovery Site in Utah.

(Courtesy of Andrew Milner)

FIGURE 4.2. (Opposite). *When the Water Becomes Shallow*

A flock of *Megapnosaurus*-like coelophysoid theropods seek new hunting grounds by swimming across a lake. When the water becomes shallow, their toes scratch the underlying sediment.

theropod dinosaurs (probably a coelophysoid like *Megapnosaurus*) were possibly struggling against the current of the lake. Large, non-swimming tracks probably made by a larger *Dilophosaurus*-like theropod were found to co-occur on the same layers of the small theropod swim tracks, suggesting that at least these larger theropods were able to wade through this area of the lake, whereas the smaller species had to swim. Based on the larger theropod footprint size, the depth of the lake at this point could be inferred to have been about 1–1.25 m. Interestingly, not only as exemplified at this site but nearly all known dinosaur swim tracks have been attributed to theropod dinosaurs, perhaps because of some type of feeding activities.

Outside of these three-toed swim tracks, other types of swimming dinosaur tracks have been reported but remain highly controversial. Among the most famous are the rather strange manus (hand) only or manus-dominated and sometimes pes (foot) only trackways that were supposedly created by swimming sauropods. Paleontologists have interpreted these tracks to mean that these sauropods were fully buoyant in water and paddled along using their arms or legs, with their hands or feet (but not both) periodically touching or "punting" off the bottom as they moved.

However, most, if not all, of these purported swim tracks have been disputed and are highly dubious. Most have been shown to have resulted from differential depth penetration of the manus or pes (due to where the dinosaur's center of mass was held, either over the legs or arms) or were the result of undertracks (a track that forms below the surface) or examples of unusual preservation. These tracks otherwise showed evidence of normal walking behavior and, although some might one day turn out to be correct, there is no convincing evidence of sauropod swimming tracks so far. This does not mean that sauropods did not swim; rather, we just do not have solid evidence for this yet.

Nevertheless, we can comfortably say that swimming was an activity that many dinosaurs routinely undertook during their lifetime, some more than others. Whether they were spending part of their lives in and around rivers, lakes, or in the sea, they would no doubt have regularly accessed bodies of water just like animals today. When it comes to tracks, I like to reiterate the point that over its lifetime a dinosaur might have left behind countless tracks and trackways but only ever one skeleton. This is why it is neat to think that some dinosaurs left behind evidence of their swimming activities, too.

THE DEATH STAR

Some of you reading this might know that I am a fan of *Star Wars* and that I sold my childhood *Star Wars* collection, among many other things, to help fund my first professional experience in paleontology (that 2008 trip to Wyoming). Ironically, there are some crossovers between fossils and *Star Wars*, including many creature designs being loosely inspired by prehistoric animals. One particular fossil has never been connected to the franchise, yet its story revolves around the death of a star. In this case, our star is a creature called a crinoid, a type of echinoderm (think starfish or sea urchin) commonly known as feather stars or sea lilies.

Stemming back nearly half a billion years, the oldest crinoid fossils are around a quarter of a billion years older than the first dinosaur fossils. You might expect that such an archaic group of creatures would have been wiped out due to the many extinction events that have happened throughout their evolutionary history, but despite being hit hard they managed to make it through. Although there were many more types of crinoids in the past, today there are more than six hundred living species found in shallow and very deep waters, some living at depths of more than 9,000 m.

Depending on where you look, crinoid fossils can be fairly easy to spot, and they have a rich fossil record, having been found at sites around the globe, but direct evidence of their behavior in the fossil record is pretty much nonexistent. Well, it was, until an astonishing discovery was made near São Bento in central Portugal at the recently discovered Cabeço da Ladeira Lagerstätte, a site of exceptional preservation, which dates to the Middle Jurassic, about 170 million years old.

This site became known as the "Jurassic Beach" and represents a prehistoric tidal flat (or mudflat), a part of the foreshore that is routinely covered and uncovered by the rise and fall of the tide, but this one recorded the development of microbial mats, layers of bacteria that helped to preserve the fossils. Located within the Serras de Aire e Candeeiros nature park, the site was discovered in an old limestone quarry and has an exposure of about 4,000 m^2 where a whole array of body fossils, especially exquisite echinoderms, and trace fossils have been found.

The trace fossils include tracks made by gastropods, shrimp, and sideways-walking crabs, some extending up to a whopping 12.3 meters long,

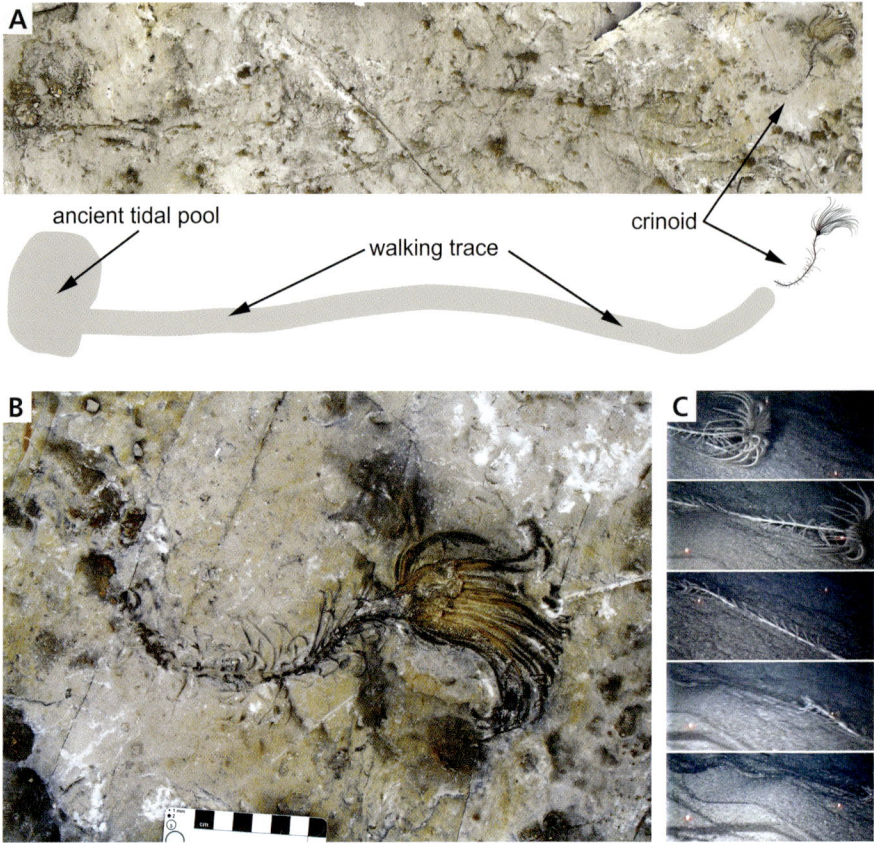

FIGURE 4.3. (A) Photograph and (B) interpretive illustration of the entire death crawl of the crinoid, including the trace (*Krinodromos bentou*) and maker preserved together. (C) A close-up of the crawling crinoid. (D) A living, crawling crinoid (isocrinid) on the seabed with a drag mark left by the crinoid's stalk.

([A–B] Courtesy of John-Paul Zonneveld and Carlos Carvalho; [C] courtesy of Carlos Carvalho; [D] courtesy of Tom Baumiller and Chuck Messing)

FIGURE 4.4. (Opposite). *Death Arrives First*

A stranded crinoid uproots itself and crawls across the sand toward the retreating ocean. Unfortunately, despite its heroic effort, it will never reach the water.

plus possible trails of fishes and insects and grazing gastropods, along with burrows made by shrimp or lobsters. However, quite unusually, preserved among the great diversity of trace fossils is a single, complete crinoid that belongs to a group known as isocrinids, which have a star-shaped stem in cross section. This, my friend, is our death star.

Many crinoids have a stalk, just like this fossil crinoid, and typically attach themselves to a substrate for most of their lives, although others are entirely free-floating and can actually swim. You might even have seen those viral videos on social media of so-called creepy crinoids walking across the seabed. These are very rare and pretty cool because it was long believed that all stalked crinoids remained fixed in one place for their entire lives. That changed, however, in the late 1980s when observations showed that some living isocrinid crinoids were able to uproot themselves and relocate by crawling with their arms, dragging the long stalk behind. You might know where I am going with this . . .

The unique fossil at Cabeço da Ladeira captures the doomed moment of a crawling crinoid baking in the Jurassic sun, caught by an abnormally low and retreating tide. Preserved directly behind the crinoid are its last moments in time, a trail showing its movements across the tidal flat in hopes of reaching the water that never came. This fossil represents a mortichnion, or death walk, capturing both trace and maker preserved together forever in a final resting place.

The exact course of the crinoid's movement can be tracked, including a series of wrinkled structures present on a patch of microbial mat that has broken sediment around it, from where the crinoid was initially anchored in a vertical hole and where it freed itself and began crawling. The start of the trail is very faint, but it becomes progressively much sharper and wider nearer to where the crinoid body lies in the matrix. A distinct and continuous central area is present along the trail and was created by the stalk being dragged behind. Two large and irregular furrows surround this central area and were produced by the crinoid's arms as it pulled itself forward. This can be observed in the actual crinoid where all of the arms and the cirri (movable appendages on the stalk that usually help to anchor and support the animal) are curving backward in a walking position. The entire death march is a little over 2 meters long.

This crawling crinoid trail is the first and only evidence of crinoid locomotion (and the only crinoid mortichnion) in the fossil record. The species of crinoid has yet to be formally identified, but the trace was recognized as something entirely new and given its own scientific name, *Krinodromos bentou*; the genus name means "the course of the [sea] lily" in Greek. In 2022, in Poland, a Triassic trace fossil was reported that might have been created by a crinoid. Though it lacks all the features of *Krinodromos*, it might be a second specimen of this specific type of trace fossil.

With such a tragic, yet incredible story, this fossil reminds me of the 9.7-meter-long horseshoe crab death trace from a Jurassic Solnhofen lagoon, which I formally described and sparked the original idea behind *Locked in Time*. Yet, in this situation, given the rarity of even witnessing a crawling crinoid on the seabed today, this truly unexpected fossil captured a sorry situation for a desperate crinoid quite literally dead in its tracks. Trying to escape from death, the crinoid was simply not fast enough to reach the water and must have experienced a slow, perhaps torturous baking-hot death one day at the "Jurassic Beach."

STAR KISSES

After flying from the UK to the United States for the first time and beginning my first-ever professional experience in paleontology, as a volunteer
at the Wyoming Dinosaur Center in 2008, I quite literally jumped at any
opportunity that came my way. In my mind, I had sacrificed so much to
make this volunteer trip possible. I had virtually bankrupted myself to get
out there, and this was my one big shot at getting onto the ladder to becoming a paleontologist. As a result, if there were any offers for extra volunteering at the weekend or invitations to go and visit a fossil site or museum,
you had better believe I was there.

One such weekend, after chatting with museum staff and volunteers, I was
invited to go fossil hunting at a local site with vast exposures of the fossil formation known as the Mowry Shale, which is famous for having an abundance
of fish remains, along with things like ammonites and rarer marine reptiles.
Formed about 98 million years ago during the Late Cretaceous, the formation
was deposited in a large inland sea that covered much of the western interior
of North America. I fondly remember finding lots of fossils on the day, especially fish scales and bits of fish jaw. After digging a bit deeper, I discovered
that the Mowry Shale also has a link to a certain famous dinosaur, *Tyrannosaurus rex*. No, *T. rex* has never been found here, but the man who discovered
the first rex skeleton, Barnum Brown, otherwise known as "Mr. Bones," also
identified, studied, and described fossils from the Mowry Shale.

Considered among the most famous dinosaur hunters ever, Brown's
most renowned discovery was *T. rex*, of course, but he had a keen interest
in fossils generally and collected thousands in his lifetime. In 1941, a short
note was published on some starlike (or stellate) fossilized imprints of
unknown origin from the Mowry Shale, and suggestions as to their identity
were requested. This brief report was written by the paleontologist Harold
Vokes, who worked at the American Museum of Natural History (AMNH),
alongside Brown. Vokes initially received a few suggestions, but he did not
deem any of them appropriate.

In November 1941, Brown had been undertaking fieldwork near Billings,
Montana, and had fortuitously obtained two large slabs of matrix from the
Mowry Shale that also contained a number of these unusual, starlike struc-
tures. Further still, the AMNH had acquired another specimen, and many

more were found at Mowry Shale localities in Wyoming and Montana. Brown teamed up with Vokes to redescribe these peculiar structures and attempt to identify their maker.

The distinctive imprint is made up of eight sharp grooves or lines typically evenly spaced and radiating from a central area, forming a starlike outline. At the end of each groove, positioned in the center, is an uplifted pile of sediment, as if an organism had placed its feet into the substrate and evenly ruffled the surface. To give you a sense of scale, the initial specimen reported by Vokes had a total diameter of 64 millimeters with each groove being around 16 millimeters long, 2–3 millimeters wide, and 3–4 millimeters deep. Both specimens Brown secured included thirteen identical impressions on a single slab, with the largest of the two slabs measuring almost 2 meters long by .5 meter wide. In both examples, eight of the stars seemed to have been made by a single individual, each star differing only slightly in diameter and length of the grooves but without a set distance between one star to the next.

After studying at least forty-eight stars, Brown and Vokes determined that no similar or comparable trace fossils are known and that they might be useful in stratigraphic studies of the Mowry Shale. Therefore, they chose to give them a scientific name, *Asterichnites octoradiatus*, in reference to the starlike shape. In terms of the maker, no identifiable fossils have been found in association with any of the impressions. It was initially suggested that the traces might be the tracks of some type of tetrapod, traces of a marine worm, or that they were made by the mouth parts of some bottom-feeding organism, but none of this added up.

Based on the shape and number (eight) of distinct grooves showing bilateral symmetry, combined with the fact that these traces were created in a marine environment, they attributed them to a squid-like cephalopod. Serendipitously, supporting their idea of a squid origin, while searching through the collections of the AMNH, they came across a squid fossil from the famous Jurassic Solnhofen limestones in Bavaria, Germany, that showed clear evidence of eight distinct arms arranged in the same relative position and with the general shape as the Mowry stars. Many more specimens from Germany have since been found with similar preservation, including those with preserved arms containing numerous hooklets and even exceptional examples with near-identical starlike arm imprints directly connected with the squid, confirming the identification.

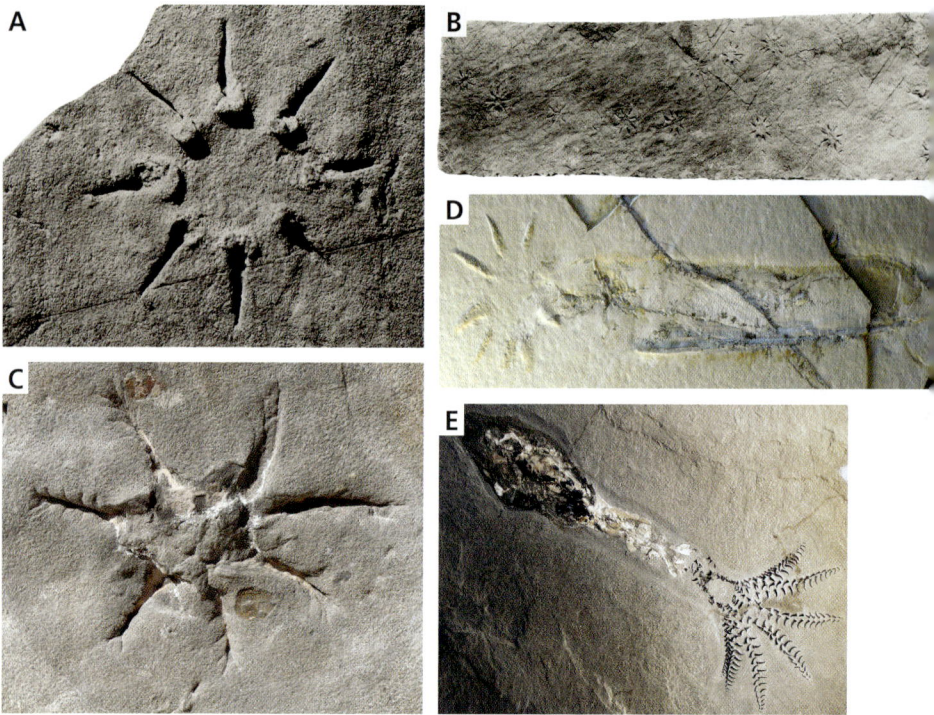

FIGURE 4.5. Some of the first starlike specimens assigned to *Asterichnites octoradiatus*, including (A) an isolated star and (B) a large slab containing thirteen individuals, from near Billings, Montana. (C) One of several specimens with hooklike impressions reported in 2019 from Natrona County, Wyoming. (D) A *Plesioteuthis* squid from the Late Jurassic Eichstätt region in Germany showing the starlike outline of its arms. (E) A squid-like *Clarkeiteuthis* from the Early Jurassic of Holzmaden, Germany, showing the starlike arm crown with hooklet-bearing arms.

([A–B] Images from B. Brown and H. E. Vokes, "Fossil Imprints of Unknown Origin: Further Information and a Possible Explanation," *American Journal of Science* 242 (1944): 656–72; reprinted by permission of the *American Journal of Science*; [C] courtesy of Melissa Connely; [D] image from D. T. Donovan and D. Fuchs, "Part M, Chapter 13: Fossilized Soft Tissues in Coleoidea," *Treatise Online* 73 (2016): 1–30; [E] courtesy of Ghedoghedo, Wikimedia Commons)

FIGURE 4.6. (Opposite). *A Star Marks the Spot*

A squad of Jurassic squid search for food in the undisturbed seafloor sediment, leaving beautiful star-shaped marks behind.

More recently, in 2019, the paleontologist Melissa Connely of Casper College, Wyoming, reported a significant collection of rare vertebrate trace fossils found in the Mowry Shale. Based on years of investigations by students and faculty, this research started in 2004 and was focused on a site located in the Bureau of Land Management's Off-Highway Vehicles Park in Natrona County, Wyoming. The key focus was to identify the vertebrate traces, which include some possibly made by marine reptiles, such as crocodilians and ichthyosaurs, but among these were more examples of starry *Asterichnites*, many of which now reside at the Tate Geological Museum at Casper College.

Often occurring as ten associated impressions, not only were many stars found around the other trace fossils, capturing an interacting ecosystem of vertebrates and invertebrates, but at least one of the stars showed grooves with hooklet-like impressions comparable to those found on the arms of squid. Thus, providing further confirmation of the maker of these unusual star-shaped traces. In actual fact, looking at some of the original stars described by Brown and Vokes, some of them also showed hooklet-like impressions directly associated with the grooves, although this association was seemingly missed and could have provided additional support for their initial interpretations.

What were these squid doing? Brown and Vokes compared the potential behavior with modern squid and proposed that the ancient cephalopod was searching for the best spot to insert its eggs, thus delivering squid "kisses" in the substrate. In modern squid, during reproductive activity, they search the sea floor for the perfect place to deposit their egg sacs, a behavior that might be undertaken communally or in isolation. Supporting such a claim, evidence of a modern squid (the longfin inshore squid) has been recorded showing a female bouncing along the seafloor with only the tips of her arms before delivering the all-important egg sac. Alternatively, these star kisses might be the result of hunting behavior, where the squid actively probed the seafloor looking for food, such as soft-bodied marine worms. This might imply that perhaps a shoal of squid engaged in this behavior together.

Whether these squid were depositing their eggs or hunting for food, their traces were preserved in fine detail resulting from rapid burial by volcanic ash that was deposited into the water, flowing over the traces. They capture the sporadic movements of ancient squid kissing the seabed, whose mystery only came to be solved through another time and another place.

HITCHING A RIDE IN STYLE

The year was 2015, and I had just finished a meeting lifting weights in the gym with a good friend and fellow paleontologist, David Penney. It beats sitting around a desk in an office. Dave and I have been friends for over a decade and share a common interest in paleontology and the gym, so we get on very well. He is also the founder of the book publisher Siri Scientific Press, which incidentally published my first two books and is where our paths initially crossed. Following our meeting, I got chatting with Dave about one of my ideas for a future book—and maybe a TV series—dedicated to fossils with extraordinary direct evidence of behavior (the book was *Locked in Time*, though it was still untitled at that point).

As a world expert in arachnids and insects, Dave quickly spurted out a flurry of ridiculously cool examples of behavior preserved in that oh-so-incredible golden amber. I quickly jotted all of this down, but one stuck in my mind, a bit like the ancient sticky resin.

The study he mentioned was one that he led back in 2012 when he was an affiliated researcher at the University of Manchester, England, where he had completed his PhD on amber in 1999. Funnily enough, he managed to obtain funding for his PhD, at least in part, off the back of the public's newfound fascination with amber due to the impact of *Jurassic Park*. Seriously. This study, however, focused on the association of two unrelated arthropods in sixteen-million-year-old Miocene Dominican amber that captured a rare and unusual association.

Amber is a unique time capsule of fossilization that can preserve ancient animals and plants with lifelike fidelity, sometimes even capturing behavioral interactions among species. The formation of amber, where sticky resin oozed over its unassuming victims, leaving them forever entombed, captures evidence of moments in time. For these reasons, amber is unique in preserving many instances of behavior and is an exciting area of study.

Contained inside this tiny piece of amber collected from the La Bucara mine in the Dominican Republic was a perfectly preserved and rare male mayfly, a type of flying insect known from more than three thousand living species. Anglers will know them well as they are commonly used as fishing bait. Mayflies, which are not only found in May, have some unusual habits and lifestyles. As aquatic nymphs, all they want to do is eat and make it through to adulthood, which takes about a year. Then, as adults, their focus switches entirely to sex and reproduction, and they do not eat. Depending on the species, some adult mayflies live for less than one hour to a few days! Mayflies and their ancestors have been fluttering around for more than 300 million years, have overcome major extinction events, and still thrive today, so maybe they are onto something here.

One thing that was definitely "onto something" was a miniscule arthropod, less than 0.3 mm long, clinging onto the base of the right forewing of this mayfly. That arthropod is a springtail, a successful group of tiny wingless arthropods that most people would easily miss given their miniature size, but several thousand species are known and are found in almost all habitats today. They have a fossil record dating back to the Early Devonian, a little over 400 million years.

The mayfly was identified as an example of a species called *Borinquena parva*, and the springtail probably represents a type of *Sphyrotheca*, a genus which is known from multiple species today. Considering the tiny size of this arthropod pair, to visualize the association in detail and to understand what was happening here, Penney and his team arranged for the piece of amber to be scanned using X-ray computed tomography (CT). This revealed the most delicate details of how the springtail had strapped itself onto the mayfly, specifically by using its prehensile antennae, a bit like how some monkeys wrap their tails around branches. The spot at the base of the wing seemed ideal, as there was a reduced risk of being dislodged during flight compared with other potential attachment points. The resulting scan of this fossil is incredibly cool, and I have included a link to this in the references, which I suggest you check out. You will not be disappointed.

The association can be explained as a type of behavior called phoresy. This is a form of symbiosis where one individual of a species catches a free ride on another, often larger species. Think of it like a free taxi service for a tiny hitchhiker. This behavior is significant for phoretic species because it plays a key role in their dispersal into new microhabitats. What is most unusual about this amber interaction is that it represents the first evidence of phoresy in a fossil or living adult mayfly and is the first record of springtails using winged insects for dispersal.

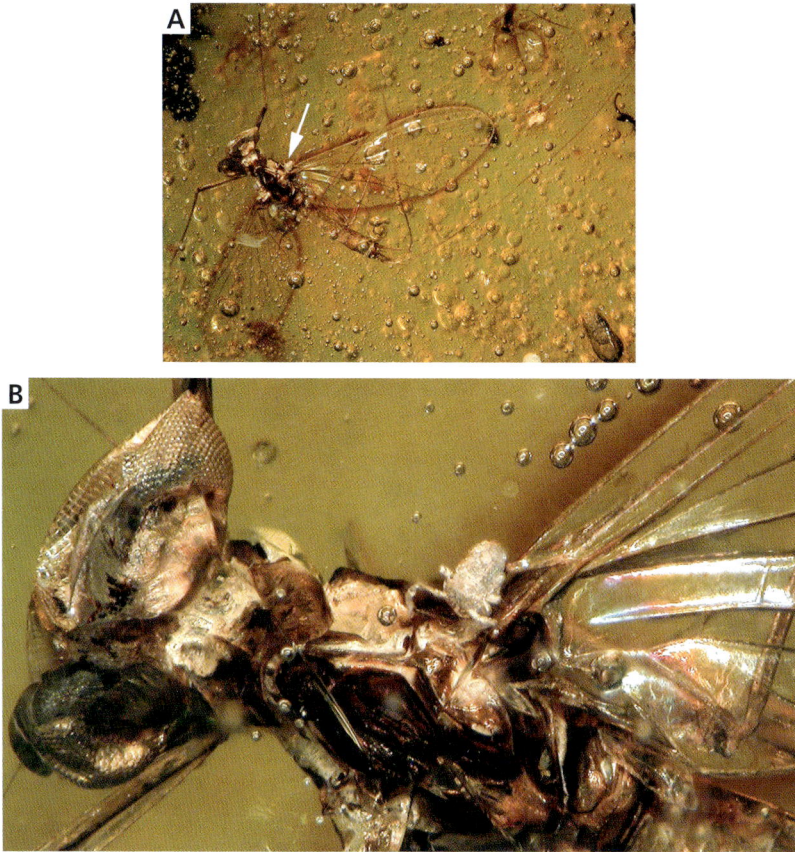

FIGURE 4.7. (A) The perfectly preserved male mayfly (*Borinquena parva*) in Dominican amber with a tiny springtail (*Sphyrotheca*) attached to its right forewing. (B) A close-up showing the hitchhiking springtail.

(Courtesy of David Penney)

Up until this discovery, it had been presumed that springtails ventured far and wide through their attachment to aerial plankton and via oceanic currents. This fossil, however, provides additional evidence for the global distribution of springtails today and indicates that this phoretic behavior may still exist among some modern species, but it has yet to be recognized. This highlights the importance of such prehistoric species in presenting new information to better understand modern counterparts.

Bob Nicholls Art

Further compelling evidence for Penney and his team's findings was reported in 2019, once again from sixteen-million-year-old Dominican amber. This time, rather than being attached to a mayfly host, a staggering twenty-five springtails were found directly attached or almost contacting the body of a winged termite and a flying ant. The winged nature of both hosts once again highlights the view that at least some springtails relied on phoresy for dispersal. The diversity of now three types of winged arthropods shows that they were not so picky about their taxi hosts.

There is probably no record of the mayfly-springtail association among living species since springtails have a distinctly nervous disposition and a unique ability to spring away from danger, using a special organ located on the underside of their abdomen called the furca. Their anxious nature, combined with their springing action and the fact that adult mayflies have an exceedingly short lifespan, is perhaps why no living species have been found hitchhiking a ride on an adult mayfly. The fact that this springtail is still attached to its symbiotic host suggests the pair were entombed instantaneously by the gloopy resin, perhaps while the mayfly was taking a break from all the sex, resting on the bark of a tree in the Dominican amber forest.

FIGURE 4.8. (Opposite). *The Doomed Stowaway*

A tiny springtail (*Sphyrotheca*) clings to the swiftly flapping wing of a mayfly (*Borinquena parva*). Their final journey will end when they both become trapped in sticky tree resin.

WATCH WHERE YOU STEP

We have all done it. Walking in the park, walking down the street, or marching into the kitchen and then *splat*. You stepped in dog poop. If it is in your house, then hopefully it was from your own dog and not your neighbor's. If you have avoided this notoriously gross situation, then good for you. Apparently, there is some superstition that says it is good luck to step in poop with your left foot and bad luck with your right foot. I disagree. It is just nasty either way. There are even websites full of stock images of people stepping into or nearly on poop. Imagine that. People have actually taken the time to photograph that sticky situation, put it up on the internet, and tried to sell it.

Much like that poop on your shoe, stick with me, as all this poop-stepping discussion has meaning and allows us to ponder whether something so familiar and gross in our world today happened in the past. And, not to disappoint, we have an unexpectedly cool fossil capturing that oh-so-nasty moment.

In 2018, a team led by the paleontologist Kazim Halaçlar undertook fieldwork at the Na Duong coal mine in Lang Son Province, Northern Vietnam, which resulted in the collection of more than a hundred coprolites. Based on their shape and structure, along with comparisons of modern crocodile feces, most of these prehistoric poops were attributed to some type of crocodilians that were living in and around an ancient Eocene lake or river around 34–39 million years ago. The coprolites are so distinct, and the scientists confident with their identification, that they decided it was possible to classify them as a new coprolite ichnospecies, *Crococopros naduongensis*. The name is derived from Latin and, rather suitably, means crocodile dung.

Nestled among this haul of fossil feces was a small, 10-centimeter-long coprolite with an unusual impression. It had the typical shape of the other croc coprolites, but something seemed off. Rather strangely, and certainly out of the ordinary for coprolites, this oval-shaped example had an unmistakable impression of two fingers pressed into it. The end of the coprolite with the fingerprints is also flattened, showing that something had stepped on it when it was still fresh and soft but firm enough to absorb the impression and ultimately fossilize. But who did the fingerprints belong to? Think

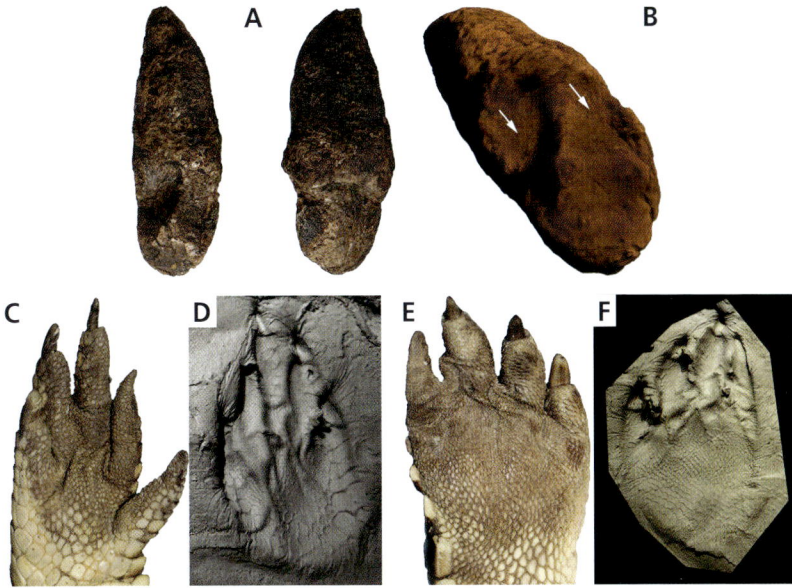

FIGURE 4.9. (A) The 10-cm-long footprint-bearing coprolite with two finger impressions. (B) 3D visualization of the impressions, showing the depth of the imprints. (C) Left manus (hand) of a Siamese crocodile and (D) a cast made of the impression left behind. (E) Right pes (foot) of a Siamese crocodile and (F) a cast made of the impression left behind.

(Courtesy of Paul Rummy)

of this a little bit like detectives looking at fingerprints at the scene of a crime, though in this case those prints are present within ancient croc poop.

Initially puzzled by this enigmatic coprolite, the team quickly ruled out mammals, turtles, and lizards and concluded that it was a croc. The decision was made after the team visited a crocodile farm in Beijing to examine numerous tracks and trackways produced by living crocodiles. They determined that the prints matched closely with finger digits IV and V of a crocodile. Peculiarly, when ruling out other possibilities, they did consider a highly unlikely scenario in which the impression was made by a crocodilian's penis as it penetrated a pooping female. Remember, crocs have an opening called a cloaca that is used for sex and to expel waste, so the team

Bob Nicholls Art

wanted to assess all possibilities considering that a croc's penis has a medial groove that might have created such an impression. Needless to say, they quickly ruled out this sexual scenario because the shape would not quite fit.

Besides the pile of croc coprolites, more than fifty croc body fossils have been found at the Na Duong site, and studies suggest that at least three different types were present in the ancient lake. Based on the size and identification of the finger impressions, each measuring 4 centimeters long by 1.5–1.3 centimeters wide, respectively, along with comparisons with fossils and modern tracks, the team proposed that a 2-meter-long croc must have stepped into the poop of another croc with its right manus (hand) or stepped into its own poop. So, in this croc's case, it would be "bad luck," though this was certainly a lucky find for the paleontologists. Further still, to give that real sense of splat, the surface texture of the coprolite bears a thin layer that the team suggest might have been made by the mucosal excretion during defecation. Nice.

As you might expect, this is the only evidence of a crocodilian footprint preserved on a crocodilian coprolite. It is amusing to say that we have evidence of a multimillion-year-old croc that stood in poop, but there is always more to it than that.

What makes this find so very rare is that we have a two-for-one fossil. Specifically, we have two different types of trace fossils preserved together, with one only existing because of the other. A trace fossil on a trace fossil. Some of the most unusual stories come from the most common behaviors yet make for some of the rarest fossils. Who would have thought that an ancient croc stepping in poop would be such a significant discovery for the Na Duong fossil site and get so many fossil fans excited over 30 million years later?

FIGURE 4.10. (Opposite). *Putting a Foot in It*

A 2-meters-long brevirostrine crocodilian walks around the edge of an Eocene lake . . . and steps on a not unsubstantial turd.

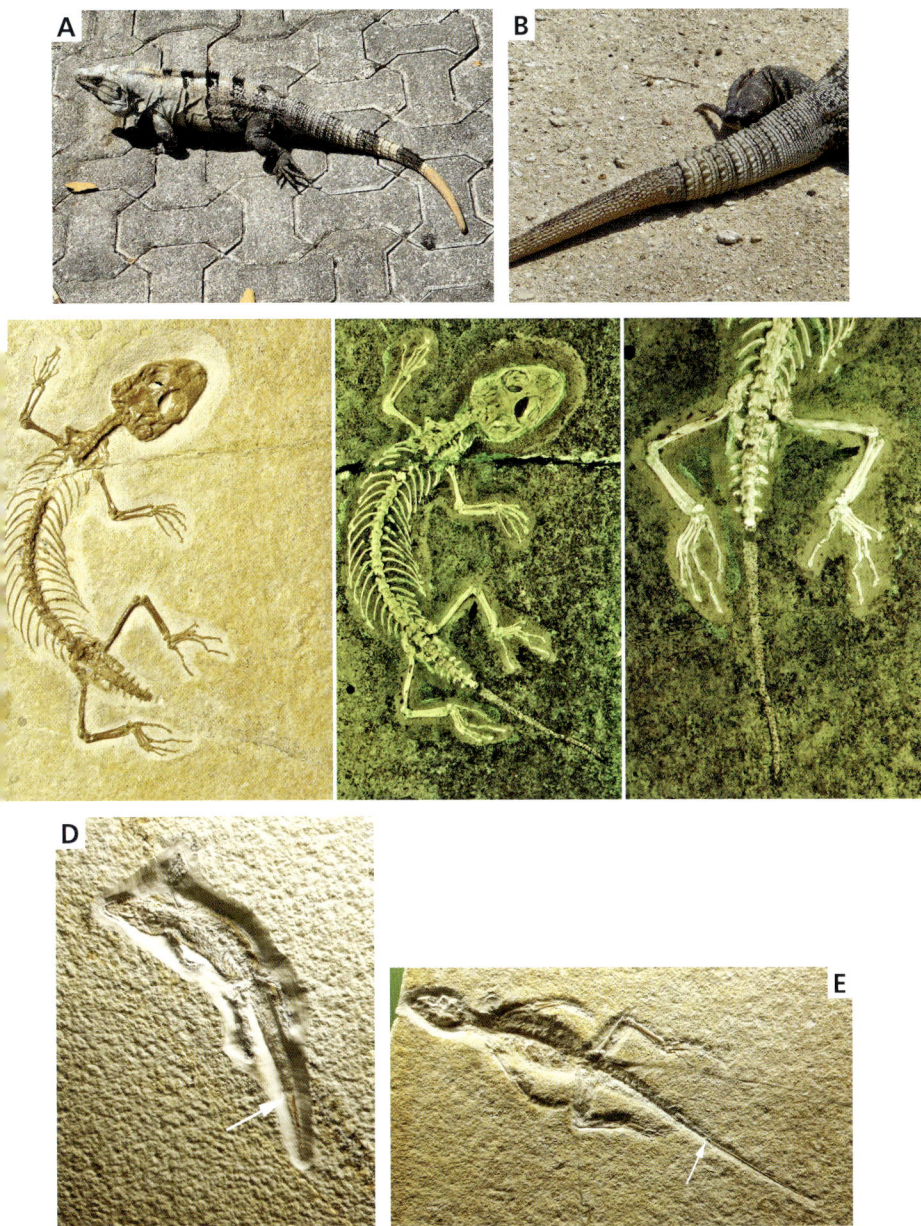

FIGURE 5.1. (A) A black iguana with a recently regrown tail. Note the difference in color and structure. (B) A close-up of another black iguana with a regenerated tail showing a clear difference in structure, pattern, and color. (C–E) Three exceptional specimens from the Solnhofen area in southern Germany. (C) The lizard *Eichstaettisaurus schroederi* with a seemingly "missing tail" that can be detected under UV light, thus revealing a regenerated, cartilaginous tail. (D) The lizard *Ardeosaurus brevipes* and (E) the rhynchocephalian *Homoeosaurus* with cartilaginous, rodlike regenerated tails.

([A–B, D–E] Photographs by the author; [C] courtesy of Helmut Tischlinger)

difficult to visualize these little theropods terrorizing their lizard prey. No doubt, having a detachable tail as an effective protective device aided these lizards, sometimes providing enough of a chance for a last-minute escape.

Another worthy inclusion here, and something radically different, is the purported Early Jurassic croc with a partly regenerated tail. The croc in question is represented by an almost complete skeleton collected from Ohmden, near the famed Holzmaden area, also in southern Germany. The tail tip appears to have been amputated, perhaps bitten off during a predation attempt, but shows signs of regeneration with the presence of a cartilaginous tip.

Modern crocodilians are incredibly hardy reptiles capable of surviving and even thriving with severe wounds, such as missing limbs and broken jaws. Although many records and studies will say that no living croc has been found with evidence of any regeneration, some extremely rare examples have been documented, including one case of tail regeneration in a black caiman that had a 21.5-centimeter-long rod of calcified cartilage. One recent study even reported evidence of tail repair and regrowth in a juvenile American alligator. Naturally, this is not the same as caudal autotomy. Still, these examples, including the Jurassic croc, show how yet another group of reptiles have extraordinary healing abilities and can overcome dramatic experiences.

Deliberately amputating your tail is an extreme yet striking evolutionary escape mechanism. Just ponder this. Over many millions of years, your ancestors developed a bony tail to aid with walking, swimming, display, or whatever it might have been, only for that tail to eventually become an expendable body part in the face of danger.

Discoveries like those observed in the captorhinids show us that some form of caudal autotomy has been occurring for almost 300 million years and has evolved in multiple different lineages, still present in a range of reptiles today. Dropping your tail as a last resort might be dramatic and a little traumatic, but this behavior can puzzle predators and increase your chances of survival. Although your bony tail may vanish, the power of regeneration will leave you with a cartilaginous tail substitute that may once again save your life.

FIGURE 5.2. (Opposite). *Living to Tell of the Tail*

The ambush was successful, but the confused *Compsognathus longipes*, a little theropod dinosaur, will have to be satisfied with a wiggling tail. Meanwhile, the tailless lizard, *Ardeosaurus brevipes*, makes a quick getaway.

DECAPOD HOUSING SUPPLIES

Many modern crustaceans live inside burrows. Whether they are marine or terrestrial species, some may live in small, temporary burrows whereas others might live individually or together in a complex maze. Such burrows serve as a refuge from predators, protection from extreme weather, a good space to keep food, or simply a spot to find a quiet moment to themselves. I can appreciate that last one.

Based on telltale signs of claw marks, along with burrow shape and size, a huge number of fossilized burrows are known to have been made by crustaceans, primarily marine decapods such as crabs, lobsters, and shrimp. Rather tormentingly, although plenty of these burrows have been found, they are almost always empty. One of the most common types of burrows thought to have been created by decapods, among a few other animals, goes by the scientific name *Thalassinoides*. This type of three-dimensional burrow is especially common and represents a type of dwelling or feeding burrow system. Quite incredibly, stemming from as far back as the Triassic through to more recent times, we have evidence of multiple *Thalassinoides* or *Thalassinoides*-like burrows with occupants still entombed inside.

Examples have been documented from around the world, from Italy to Slovakia and from the United Kingdom to the United States. Often, these *Thalassinoides* burrows are made up of a complex network of systems and include the crustacean(s) sitting snug inside. Some intriguing examples include burrows from Early Jurassic rocks of East Greenland that contained well-preserved individuals of a long-tailed lobster, *Glyphea rosenkrantzi*, along with microcoprolites inside the burrows. At an Early Cretaceous site in Portugal, within the Boca do Chapim Formation, an extensive area was found with numerous burrows containing successive populations of the lobster-like crustacean *Mecochirus rapax*. A rare find from the Eocene rocks of Mount Discovery, East Antarctica, yielded the first evidence on the continent of fossil ghost shrimp, *Callichirus symmetricus*, inside their burrows. In southeastern Australia, multiple burrows from Miocene-Pliocene deposits were found with *Ommatocarcinus corioensis* crabs inside. So, it is safe to say that we have a good collection of these types of burrows and occupants, each of which just happen to have difficult to pronounce names.

But these denizen decapods did not simply live inside burrows; they also took up residence inside the shells of cephalopod mollusks, especially ammonites. A study published in 1995 focused on four ammonite specimens from the Jurassic of England and Germany. Three of the four ammonites came from the Late Jurassic of the Isle of Portland, southern England, and belonged to the perisphinctid family; all three contained the same type of lobster, *Eryma dutertrei*, whereas the single German specimen came from older Early Jurassic rocks (~180 million years old) from the famous Posidonia Shales of Dotternhausen, southern Germany, and consisted of a complete macroconch (a female) *Harpoceras* ammonite containing a single lobster, probably representing a type called *Palaeastacus*. This latter lobster is well-preserved, almost complete, and sits in the front half of the ammonite chamber, with its head pointing toward the ammonite entryway, as if saying hello. Some small, spherical microcoprolites were also preserved inside the chamber, suggesting that the lobster lived there for at least some time.

In 2012, the description of a most intriguing specimen of yet another Early Jurassic *Harpoceras* from Dotternhausen was found to contain three delicate lobsters inside the living chamber. All three lobsters measure just over a couple of centimeters long, belong to the same type, and can be assigned to an extinct family called Eryonidae. They are oriented in the same way (dorsal side up), closely spaced, and positioned with their tails angled together. As with the other specimens, there is no doubt that the lobsters are preserved inside the ammonite's chamber, rather than atop or under it.

Several undescribed Late Jurassic examples are also on display at The Etches Collection in Kimmeridge, Dorset. But the association is not just restricted to the Jurassic, as some Late Cretaceous crabs from France have been found in the body chambers of ammonites, too. A single, small crab was even found inside a baculite (straightened ammonite) from the Late Cretaceous of South Dakota. In 2020, a Late Cretaceous ammonite from the Mangyshlak Mountains of western Kazakhstan was also found to contain a complete lobster lodged within the body chamber of an ammonite. In that case, the lobster was distinct enough that it turned out to be an entirely new species, *Hoploparia tectumque*. All of these lobsters and crabs were certainly living it up in their cozy shelly homes, at least temporarily, but what were they doing?

First, none of these examples represents the last meal of the ammonite. Each decapod is too complete, preserved literally inside the living chamber (and not the gut region), and most are too big to have been swallowed whole. Rather, the type of association observed in these fossils is referred to as *inquilinism*, where one animal lives in or exploits the living space of another, using the dead or alive host as a home. We know from the fossil record that Mesozoic seafloors must have been littered with empty ammonite shells. Clearly, they did not go to waste. They offered shelter and provided these decapods with a space to live and hide inside, perhaps using them as personal dining rooms.

However, as the various researchers who have studied these associations proposed, a key reason the decapods likely exploited the lifeless shells was for molting. As we know from modern decapods, they are most vulnerable during and immediately after molting, so the shell would provide safety. Some of the fossil decapods were found to be molts whereas others were the actual carcass, so we know that they lived at least part of their lives inside these dead shells. Similar evidence has also been recorded in trilobites that have been found inside the empty shells of much earlier cephalopods.

The ammonite containing the three lobsters is worth a special mention. This fossil is the only example with multiple decapods, showing that some prehistoric species spent time together in groups, meaning they were likely gregarious. In this case, the three specimens represent corpses rather than molts, and they clearly entered the ammonite shell together for use as some kind of shelter. For what specific reason is difficult to say, but as the researchers Adiël Klompmaker and René Fraaije suggested, perhaps they were preparing to molt, wanted to hide from predatory fish, were attracted by decaying soft tissues inside the ammonite, or chose this large ammonite as a long-term base on the muddy bottom. In any case, this fossil represents the oldest definite example of gregarious behavior in decapods, which is observed in many species today.

Taking their home a literal step further, we also have fossilized hermit crabs. You might be familiar with these specialized decapods that famously use empty shells, typically of marine snails, to cover their soft, squishy, and unprotected abdomens. As the hermit ages, it outgrows each shell and ditches it in favor of another, larger shell, a process that recurs throughout its lifetime. When selecting a shell, hermits can be quite picky; it must have the right shape and size, durability, and weight, and it must also have a loose enough fit at the opening so that it does not get stuck.

A few rare fossil hermit crabs have been found inside snail shells, which is what we expect based on living species. Rather marvelously, we also have incredibly rare hermit crabs that used ammonite shells as homes! How cool is that? The first specimen to be discovered and described is Early Cretaceous in age and comes from Speeton on the Yorkshire Coast in England, a place where I used to collect fossils in my teen years. The ammonite shell belongs to a species called *Simbirskites gottschei*, and positioned right at the opening of the ammonite chamber is the little hermit crab, with its pincer-like claws (called chelae) poking out; the left claw is much larger than the right. As part of the research published in 2003, this hermit was deemed to be a new species by the researcher, who happened to be René Fraaije, and was called *Palaeopagurus vandenengeli*.

A second specimen of a hermit with its attached ammonite house was described in 2006. This specimen comes from the Early Jurassic of Banz in southern Germany and includes a partial hermit, also representing a new species called *Schobertella simonsenetlangi*, poking its claws out of the opening of an ammonite called *Pleuroceras*. This one represents the oldest definite example of a hermit crab inside its shell. An incomplete hermit was found inside a fragmentary ammonite, *Craspedites nekrassovi*, from the Late Jurassic of Moscow, Russia, and may also represent another example in its ammonite house. Findings presented in this 2020 study suggested that these hermit crabs might have actively hunted the ammonites for their shells, as many of the ammonite shells bear evidence of injuries (some healed) that may have been caused by the hermits. This would be rather unusual and surprising considering that most living hermits do not hunt and kill for shells, though there is rare evidence that at least some larger-bodied hermits do occasionally kill snails and later take their shells for themselves.

Besides those examples, a few hermits have also been recorded inside much larger ammonite shells that were certainly not their mobile homes.

FIGURE 5.3 (A) A complete lobster (*Hoploparia tectumque*) nestled inside the body chamber of an ammonite (*Schloenbachia*) from the Late Cretaceous in western Kazakhstan. (B) The hermit crab, *Palaeopagurus vandenengeli*, can be seen poking out of the opening of a *Simbirskites gottschei* ammonite from the Early Cretaceous in Speeton, Yorkshire, England. (C) A modern hermit crab.

([A–B] Courtesy of René Fraaije, Oertijdmuseum, Boxtel, the Netherlands; [C] courtesy of USFWS)

Rather, evidence suggests those individuals resided inside the shells a little like the lobsters and crabs mentioned earlier, perhaps using them as temporary shelters. As an interesting side note to all of this, some modern hermit crabs, such as the Caribbean hermit crab, have been observed to use actual fossil snail shells as homes. It must be strange seeing these extinct shells given new life.

FIGURE 5.4 (Opposite). *The Squatter's Defense*

A Cretaceous hermit crab, *Palaeopagurus vandenengeli*, has made its home inside the empty shell of a long-dead ammonite. While some tiny lobsters dart for cover, our hermit crab prepares to defend itself from an approaching octopod.

Combining all of these occurrences, it is pretty cool to look at the evidence and behaviors we see today in living decapods and retrace their roots to their prehistoric relatives. The use of ammonite shells must have been common on the typical fine-grained seafloors of the Mesozoic. Although the ammonites might have gone extinct, the behavioral instinct to find a quiet spot or an empty shell for refuge is a trait that decapods continue to utilize today.

As a final thought, it is intriguing to ponder how many museums and collectors across the world might have an ammonite fossil sitting in their collections right now, currently concealing a little decapod inside its shell. I literally have a couple of ammonites in my collection behind me as I type, but I do not think I will be splitting them open anytime soon. Well, probably not.

TINY HOMES FOR TINY AMPHIBIANS

Frogs, toads, newts, salamanders, and caecilians are all amphibians. Today, there are a little over 8,500 species, with frogs making up around 90 percent. Being able to switch between worlds, comfortably living both in the water and on land, amphibians appear to have a superpower, but they also have a crucial weakness. Due to their moist, permeable skin, which helps them to breathe and absorb water, it can also leave them highly vulnerable to dehydration and desiccation in extreme temperatures.

To combat hostile conditions, amphibians have developed various mechanisms to overcome such challenges. One of these is burrowing, or at least burrow stealing or sharing. Many types use burrows opportunistically, seeking out old, abandoned burrows or even sharing shelters with different types of animals. Others might simply hide deep under the leaf litter, under log piles, or dig their own underground burrows as a safety refuge.

The amphibian story began more than 300 million years ago and one behavior they developed early was burrowing. Ancient amphibians and their kin comprised various (and now extinct) groups and families, including one fully aquatic group known as lysorophians, whose members had small heads, long bodies, and tiny limbs; outwardly they resembled caecilians or small snakes. At a site in eastern Kansas, multiple examples of a small type of lysorophian, called *Brachydectes elongatus* (or *B. newberryi* according to some researchers), were discovered inside burrows.

Known as the Eskridge locality, near the town of Eskridge in Wabaunsee County, Kansas, this fossil site is Early Permian in age, approximately 290 million years old. Burrows and other fossils, including fish, amphibians, reptiles, algae, and seed shrimp occur in a specific layer of green mudstone in what is known as the Speiser Shale. Within this mudstone layer, the burrows are found to occur vertically at three distinct levels that are separated in time, with each interpreted to represent a seasonal and shallow temporary body of water, otherwise known as an ephemeral pond (or vernal pool), that was present on a coastal plain.

The burrows cluster together in large accumulations of up to forty-five burrows. There are two types: an elongated elliptical tube 4–32 centimeters long and 2–7 centimeters wide and a much shorter elliptical tube just

Bob Nicholls Art

1.5–5 centimeters long and 2.5–8 centimeters wide. Contained inside many of these burrows are well-preserved, often coiled, and complete or fragmentary skeletons of *Brachydectes*. In complete specimens, their heads tend to point up and their tails down; some skeletons are found outside of the burrows, too. The largest individuals have a maximum length of around 50 centimeters and were found to be tightly coiled and snug inside their burrows.

The ancient critters created the burrows using their shovel-shaped snouts and tough, tiny heads. The shape and structure of the burrows are so distinct that they can easily be distinguished from the burrows of lungfish, which often occur in the same age of Permian rocks. The team that described the *Brachydectes*-burrow associations also decided to name the type of burrow, which they called *Torridorefugium eskridgensis*.

Having so many similar burrows in the same spot, each often containing the complete remains of the same type of early amphibian and recurring at least three distinct times, shows that a specific type of behavior and sequence of events must have occurred periodically.

In this case, the burrows are the result of *Brachydectes* estivating en masse, essentially entering a deep sleep (known as torpor) similar to hibernation that was undertaken during extended hot and dry periods. The mudstone beds that the burrows and other fossils are within are capped by mud cracks, indicating the drying of the pond. The Eskridge locality was not a one-off either because additional examples of *Brachydectes* found inside their burrows have previously been documented in similarly aged rocks in central Kansas, Texas, and Oklahoma, which were also found to have inhabited ephemeral ponds.

Burrowing offers protection from adverse environmental conditions and in some cases is the only option for animals to escape the extremes of weather. The sort of group estivating behavior observed in *Brachydectes* is

FIGURE 5.5. (Opposite). *A Shelter Becomes a Tomb*

The little Permian amphibian *Brachydectes elongatus* is hidden away from the seasonal drought inside its damp burrow. Unfortunately, on this occasion, the soil will dry deeper than usual and our *Brachydectes* will remain in the ground forever.

analogous with some living estivating amphibians. These include aquatic salamanders that inhabit ephemeral rivers and ponds in the southeastern United States; the critters burrow into the muddy floor where they remain inside hardened mud until the heavy rains return. In more extreme cases, desert anurans like Couch's spadefoot toad and African bullfrog are only active for two to five months of the year and spend the rest of their time in estivation burrows.

As the small pond began to dry, these lysorophians burrowed into the bottom for apparent safety, but this would be their downfall—literally. This burrowing behavior might have been in response to episodic, and perhaps seasonal, droughts that dried out the pond for longer than the individuals were able to last. Unfortunately for these tiny tots, the drought was so extreme that it baked them inside their burrows where they remained for all eternity.

UNDERCOVER INSECT

Some animals have evolved extreme adaptations to evade being spotted. Camouflaged and blended into their environment, appearing seemingly invisible increases their chances of survival. If you are a predator, this adaptation can also improve your chance of sneaking up close enough to your prey without being seen. Taking this up a notch, various species have evolved the ability to mimic and closely resemble entirely unrelated species or even inanimate objects.

Mimicry is a fascinating behavioral adaptation. Perhaps my favorite example is the spider-tailed horned viper from Iran that expertly wields its spiderlike tail as bait. Concealed in its environment, this snake stays perfectly still except for its tail, which shimmies around like a spider, luring unsuspecting birds to swoop down for an easy meal only for the snake to strike. It is a pretty fabulous example of mimicry, though not so great for the birds. Another one that blows my mind, and a bit of a reverse of the other, is the snake mimic caterpillar, found in the Amazon; it mimics the head of a venomous snake, complete with apparent eyes. It is also able to perform a "strike-like" movement to deter potential predators.

Undoubtedly, some of the most astonishing masters of disguise are insects that have evolved to resemble leaves. Yes, leaves. Can you imagine the first time somebody witnessed one of these crawling around? Being able to hide in plain sight is the ultimate form of camouflage, with some species imitating not only the shape, structure, and color of a leaf but even the way that it moves, just like a leaf swaying in the wind. Some types look as if they have bite marks on them, others have body shapes that look like the thorns of a plant, and some look like flowers. Understanding how complex mimicry can be in modern animals, evidence in fossils might seem highly unlikely, yet insect mimicry of plants stretches back an astonishing 270 million years to the time of the sail-backed *Dimetrodon*, the Permian Period.

The fossil in question is what is known as a katydid, or bush cricket, a type of insect that belongs to a large group called Tettigoniidae, which contains several living and extinct families found on all continents except Antarctica. Living species are well-known for their leafy-like bodies and spectacular plant mimicry, being perfectly camouflaged in their modern-day habitats.

FIGURE 5.6. (A) The common true katydid doing a perfect job of imitating a leaf. (B) Photograph and (C) interpretive illustration of a stick insect from the Early Cretaceous in the Araripe Basin, Brazil.

([A] Courtesy of James St. John, Wikimedia Commons; [B–C] courtesy of Victor Ghirotto, et al. "The Oldest Euphasmatodea (Insecta, Phasmatodea): Modern Morphology in an Early Cretaceous Stick Insect Fossil from the Crato Formation of Brazil," *Papers in Palaeontology* 8 (2022): e1437)

The long-extinct katydid was identified as a new genus and species, called *Permotettigonia gallica*, and was collected in the Roua Valley near the small, beautiful village of Daluis, Alpes Maritimes, in southeastern France.

The single specimen represents an isolated forewing, which is pretty typical of many fossil insects whose wings are sometimes the only parts that preserve. The nearly complete and very broad wing displays the same shape and specific venation type we see today in modern katydids, which

FIGURE 5.7. (Opposite). *Move Along, Nothing to See Here!*

A group of katydid insects, *Permotettigonia gallica*, mimic the shape of *Taeniopteris* leaves to hide from passing predators, such as this lizard-like reptile, *Bolosaurus*.

BOB NICHOLLS ART

have leaflike wings, so there is no mistaking this for an actual leaf. The wing is 2.7 times longer than wide, especially broad compared to its total length, which is estimated to be just 4 centimeters. Based on the structure of the wing, the researchers determined that the wings would be positioned in a sort of upright direction like in the common true katydid *Pterophylla*, found in North America today. It is difficult to determine what specific leaf this katydid mimicked, but it might have been a common type known as *Taeniopteris*, or a related species, which had a comparable structure to the wing.

This discovery also showed that the katydid group is much older than previously thought, as it was assumed to have been no older than the Jurassic. Prior to the announcement of *Permotettigonia* in 2016, a few examples of plant mimicry were reported from the Mesozoic and Cenozoic. Some of the best examples come from the Jurassic of China and include a hangingfly whose wings near perfectly mimicked the leaves of a ginkgo plant, lacewings whose wings imitate the leaves of cycad and cycad-like plants, and, although not a plant, another lacewing had a wing pattern that matched perfectly with a lichen.

Some of you will no doubt be wondering about stick insects, too, which are excellent disguise artists. This might be the best form of camouflage because, hey, who *really* wants to eat a stick? Just ask *Dilophosaurus* (if you know, you know). We do have evidence of fossil stick insects, or stick mimicking, with the oldest possible occurrence of sticklike insects from the Permian, although complete specimens of more modern, definite stick insects are known from the Early Cretaceous of Brazil.

Something a bit different but along the same lines, distinct coloration or patterning has been found on the wings of a range of various fossil insects and is a very ancient feature going all the way back to the Carboniferous. These spots and bands are likely to have evolved due to the increase in predators and were used as a sort of antipredator camouflage, among other things. For example, during the Jurassic and Cretaceous, there was a surge in the number of dragonflies with various patterns. Researchers have proposed that this is directly linked to the evolution of insectivorous flying vertebrates, especially pterosaurs, but also the appearance of birds and small feathered nonavian dinosaurs.

Insects imitating plants to help them carefully slip under the noses of animals is a spectacular adaptation. Knowing that this form of mimicry extends back almost 300 million years suggests that predation pressure was high enough back then that certain insects developed leaf mimicry as a form of defensive strategy—in this case, probably to avoid detection from predators like pigeon-sized griffenflies (extinct relatives of dragonflies), which have been found in the same rock formation. Just like today, mimicry must have played a vital role in increasing the chances of survival and helping such insects to evade being eaten (at least some of the time) in the very deep past.

6 | Finding Food

The giant palm salamander found in Central America has the fastest tongue in the animal kingdom. Shooting out its tongue in approximately seven milliseconds, fifty times faster than a human can blink, its prey disappears in an instant.

I GOT YOUR CHICKEN LEGS

Gazing at a towering *Tyrannosaurus rex*, observing its massive skull, gulping at its banana-sized teeth, and imagining that deadly, bone-crunching bite—it is easy to see why *T. rex* is the most studied dinosaur in the world. I often have students or members of the public tell me, "I want to study *T. rex* because it's so cool!" I usually reply, "Well, unless you can find one that turns out to be pink, has two heads, or still has its last supper in its belly, then good luck." I do this a bit tongue-in-cheek because every inch of *T. rex* has been studied to extinction, so to make a name for yourself in the rex world you must either find something radically different or reinvent the wheel and apply some new techniques or approaches to studying its bones. If *T. rex* is your ultimate goal, then the next best thing, as it were, is to keep it in the family and look at other members of the Tyrannosauridae and their kin.

The wider family of tyrannosaurs, or, more correctly, tyrannosauroids, extends back to the Middle Jurassic, more than 160 million years ago. The oldest and most primitive member of this group comes from the United Kingdom and is the labrador-sized *Proceratosaurus*, known from a single skull described in 1910; I have examined this specimen and even have a 3D print sitting in my office. It is fascinating to think that great-grandparent rex was this small theropod that would have been living in the shadows of megalosaurs and more but also that *T. rex* is closer to you and me in time than it was to its many times great-grandparent. My point is that this successful group of predators, which bowed out at the very end of the dinosaur reign with *T. rex*, has a long history filled with numerous small to medium and gigantic members, each fascinating in their own right.

There has been a vast amount of research dedicated to investigating tyrannosaur diets, with a focus on teeth and bite force, as well as evidence of bite marks and feeding traces. Although the latter has been reported among some specimens, principally in proposed predator-prey scenarios, everything is inferred, and the one key, crucial piece of evidence that has been missing is stomach contents. So far, the only direct, unequivocal evidence yet comes from an albeit smaller, but still enormous and terrifying tyrannosaur, *Gorgosaurus libratus*.

Gorgosaurus is a name that might be familiar to you. Although it has not received as much limelight as its larger cousin, Gorgo was a formidable beast 8 to 9 meters long and around 2– 3 tons. Sharing its environment with other contemporary tyrannosaurs, it was one of the top hunters during its day and age, some 75 million years ago in the Late Cretaceous, and like Big T, it also had sharp, thick, heavily serrated steak knife–like teeth perfectly adapted to tear through flesh and crunch through bone. As you might recall from our previous trip to Egg Mountain, in 2009, I helped to excavate part of a *Gorgosaurus* in Montana. This always stuck with me as one of my fondest dino digs, namely because my good friend, Greg Willson, and I found part of a Gorgo jaw with several teeth. It was one of those extreme wow moments. The specimen was eventually cleaned and displayed at the Wyoming Dinosaur Center.

Several gorgeous *Gorgosaurus* skeletons have been discovered in the badlands of the aptly named Dinosaur Provincial Park in Alberta, Canada, with the very best on display and in the collections at the Royal Tyrrell Museum in Drumheller. Growing up, watching every dino doc that I could, there was often some reference to the Tyrrell, usually discussing a recent dino discovery or new research. It was one of those museums that, to a dino-loving kid, just seemed like the coolest place on Earth. In 2017, I finally got to visit as a researcher to study dinosaurs and ichthyosaurs, and that visit did not disappoint. Walking into the collections room, I found myself greeted by an almost perfect skeleton of *Gorgosaurus* sitting there, staring at me with a big toothy grin. Walking through the galleries, I got to see another beautiful, almost complete Gorgo, one which I'd go as far as to say is a contender for my favorite theropod fossil ever (yeah, this is very niche but we paleontologists can be super nerdy about these things). One other intriguing Gorgo skeleton was being studied by a group of researchers.

This specimen was found in 2009 by the Tyrrell paleontologist and senior technician Darren Tanke, who also did a magnificent job cleaning the fossil. Tanke has made many notable contributions to paleontology, in the form of discovering amazing finds, carefully cleaning countless fossils, and writing numerous academic papers. He is the museum's longest-serving employee and is a treasure trove of information. Naturally, the perfect person to clean this *Gorgosaurus* skeleton. While carefully revealing the bones layer by layer, his keen eye spotted a strange mass within the rib cage, including some small toe bones sticking out. Had he just found this tyrannosaur's last meal?

The toes were connected to the exceptionally preserved, complete, and mainly articulated legs and feet of not one but two young oviraptorosaur dinosaurs inside the abdominal cavity. Remains of this birdlike dinosaur (called *Citipes elegans*) are rare, and an assessment of the bones showed that they each represent yearlings. They were about the size of a large chicken and had an estimated body mass of 9 to 12 kilograms.

Strangely enough, each baby *Citipes* is composed almost entirely of its legs with one of them including a few bits of the tail and some claws from the hands. Highlighting the rarity, these represent the most complete fossils of this dinosaur ever found, all thanks to having been eaten—talk about luck. A closer inspection of the legs showed that the prey was torn apart and consumed in two separate feeding events, with one pair of legs having been digested more than the other—with further evidence of acid damage on the bones—suggesting that it was probably eaten several hours to even days earlier.

This means that the *Citipes* meals were not a coincidence. Rather, this *Gorgosaurus* must have had a knack for chowing down on *Citipes* drumsticks, having a second helping of the same "chicken legs" at another time. This might allow us to infer that these little dinosaurs were the preferred prey of juvenile Gorgos. In this instance, considering that the legs were still well preserved and articulated indicates that this *Gorgosaurus* must have died not too long after consuming its prey, perhaps less than one week. For what reason is unclear.

Given that chickens are theropods and in the grand scheme have a similar leg morphology, chicken legs are an apt analogy for this pair of fossil drumsticks. But why would this Gorgo only eat the hindquarters of this animal? For one thing, anybody who has eaten chicken drumsticks will tell you that there is plenty of meat attached to those bones. The *Gorgosaurus* probably only wanted to eat the meatiest parts of the *Citipes*!

The other crucial, and perhaps most significant, part of this discovery is that this *Gorgosaurus* was a juvenile. At about 4 meters long, a little under half the size of known adults, estimates of this individual's age put it at around five to seven years old and weighing in at just 335 kilograms. It was several years away from reaching adulthood. Compared with the adults, the juveniles were much more lightly built, with long, skinny legs, and they had not yet developed those banana-sized teeth, having blade-like chompers instead.

The fact that a youngster ate chicken-sized dinosaurs as its last meal informs us greatly about prey choices and food web dynamics. Or, as the researchers put it, this discovery reveals an ontogenetic dietary shift in *Gorgosaurus*. That is a sciencey way of saying that this provides new clues as to how tyrannosaur predatory behavior transformed as the animal transitioned from juvenile to adult.

This was the core focus of the findings presented by the team of researchers who described this exceptional specimen in a 2023 study, and it confirmed what paleontologists had long suspected. Young tyrannosaurs, in this case a juvenile *Gorgosaurus*, hunted small prey but shifted their focus to larger game, think herds of big hadrosaurs and ceratopsians, as they got older, with their much bulkier bodies requiring more calories. It would be unexpected for a young, inexperienced, and more lightly built tyrannosaur to attempt taking on an adult hadrosaur or ceratopsian, which could probably crush the juvenile with its far heftier body. A good comparison can be made here with modern crocs: juveniles eat smaller fish, crustaceans, and frogs, among other critters, but adults take down far larger prey, including big mammals.

So, you would not expect this teenage tyrannosaur to just start hunting and eating big herbivores. Such differences in diet show that juveniles and adults were not in competition for prey, which probably resulted in fewer confrontations. For the youngster, it had to work its way up that theropod pecking order before it could fill the ecological niche of top predator.

Speaking of top predator status, it would be remiss not to mention a specimen of another tyrannosaur, and a contemporary of Gorgo, that was also reported with some sort of stomach contents. This time the tyrannosaur was an adult *Daspletosaurus*, collected from Teton County, Montana. Described in 2001, the fragmentary, disarticulated skeleton was found with associated, scattered remains of juvenile hadrosaurs, including four acid-etched vertebrae and part of a jaw. Estimates suggest the vertebrae belonged

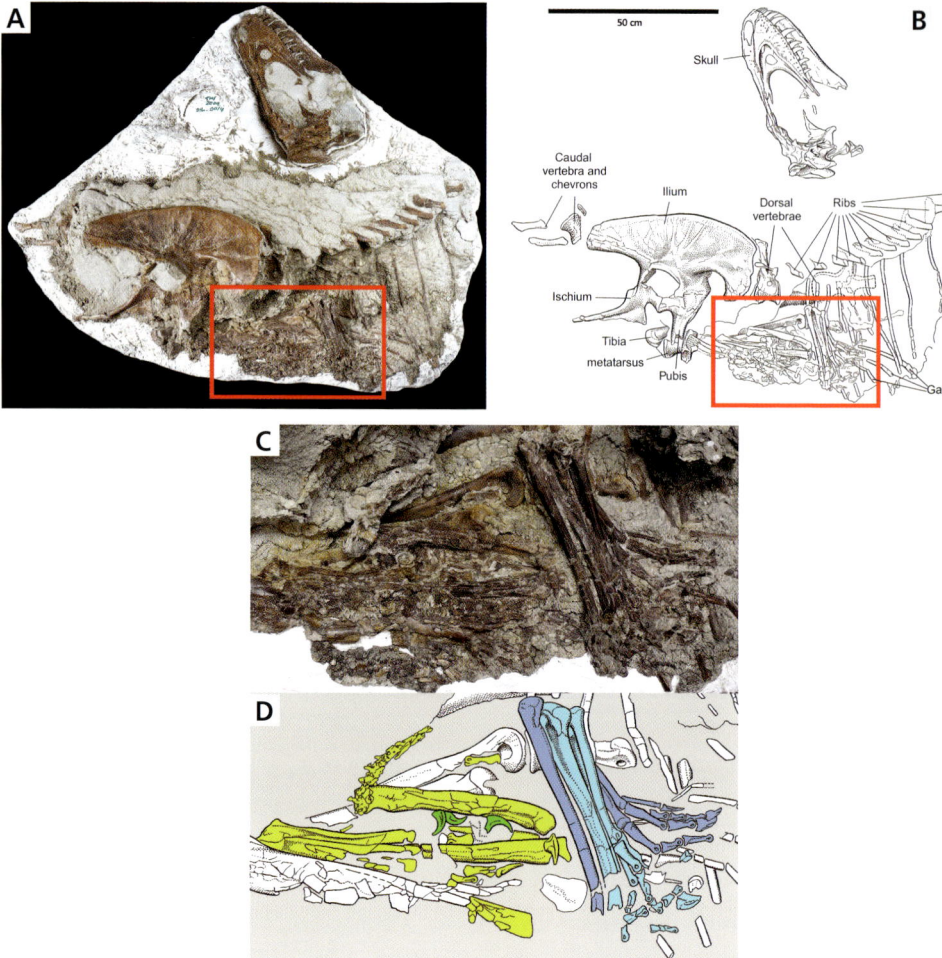

FIGURE 6.1. (A) Photograph and (B) interpretive illustration of a juvenile *Gorgosaurus libratus* with its last meal of *Citipes elegans* "drumsticks" inside its stomach. (C) Photograph and (D) interpretive illustration of a close-up of the stomach contents. Light blue and dark blue represent the hind limbs of the anteriormost *Citipes*, whereas light green and dark green represent the more posterior *Citipes* individual.

(Courtesy of François Therrien)

FIGURE 6.2. (Opposite). *Only the Best Will Do*

A long-legged juvenile *Gorgosaurus libratus*, a tyrannosaurid theropod, tears off a meaty leg from his prize, a feathered *Citipes elegans* (a caenagnathid theropod).

BOB NICHOLLS ART

to an individual about 3 meters in length, whereas the portion of jaw might have come from an animal with a body length of about 1–1.2 meters; both are much larger, in terms of weight and length, than the *Citipes* babies consumed by the juvenile *Gorgosaurus*. The evidence of stomach contents in this *Daspletosaurus* is much more ambiguous, which is why the *Gorgosaurus* specimen is so exceptional—its last meal is nestled right where it would be expected, inside the gut of this tyrannosaur.

Playing with toy dinosaurs as children taught us that dinosaurs ate dinosaurs. It's the same fact we learned through TV documentaries and movies, though the reality of the fossil record is that only a smattering of specimens provide us with extraordinary direct evidence. At present, we might have discovered somewhere in the region of about twenty-five carnivorous theropods with their last meals. Staggering, when you think about it, considering the enormous numbers of dinosaur remains worldwide.

This is what makes discoveries like our chicken drumstick–eating teen tyrant so significant. They help reaffirm our thoughts and ideas regarding research, present us with new facts and figures that change and challenge our understanding, and offer us unique glimpses into the lives of creatures long gone but never forgotten.

A NOT SO WHALE OF A TIME

At around 8 meters long, the orca is one of the largest apex predators alive. You will undoubtedly know it by its other name: the killer whale. This name can confuse some people because killer whales are actually dolphins, which belong to a group of toothed whales (known as odontocetes). The name game can get a bit confusing, I know, but essentially all dolphins are whales but not all whales are dolphins.

The name "killer whale" originates from sailors who witnessed orcas killing other types of whales, so the name makes sense. Plus, if you have seen any nature documentary starring orcas, you will almost certainly have witnessed some impressive, often cooperative hunting behaviors with orca pods taking down an array of animals. Well known for their large size, intelligence, and sociality, they can even kill great white sharks and significantly larger sperm whales and blue whales. Some might say they are the ultimate apex predator alive.

From 50-million-year-old wolf-like walking whales to gigantic blue whales (the largest animal on the planet), the evolution of whales, or cetaceans, is quite remarkable. The fossil record has provided us with some exceptional specimens that have helped piece together their incredible evolutionary journey, whose story only truly began following the demise of the mighty marine reptiles at the end of the Cretaceous Period. With that said, it took several million years before early whales finally took the plunge and fully committed to life at sea. Even then, it was not until about 40 million years ago, during the Eocene, that the first truly gigantic whales appeared, such as *Basilosaurus*.

Let me tell you, *Basilosaurus* was a beast. At about 15–20 meters in length, with large teeth and bone-breaking bite force, it was the first massive predator to appear in the oceans since the disappearance of mosasaurs and plesiosaurs; it still retained hind limbs although they were not used for locomotion. When it was first described in 1834, it was thought to be some enormous reptile, hence the "saurus" part of the name, but it was later revealed to be a type of early cetacean and specifically belonged to a now-extinct group of early cetaceans known as archaeocetes, which are the ancestors of today's whales. Two species of *Basilosaurus* are identified today,

B. isis and the slightly geologically younger *B. cetoides*, and they belong to a family of archaeocetes fittingly named Basilosauridae, which contains multiple subfamilies.

As an apex predator, it would be expected that *Basilosaurus* could have fed on practically anything, much like a modern-day orca. We know that orcas are famed for feasting on whales, though they dine on other marine mammals, such as seals, sea lions, and walruses, and we also know that they eat fish, squid, turtles, and seabirds. Anything goes for the orca. As for *Basilosaurus*, which was about three times the length of an average orca, it needed plenty of calories to fuel its heavier, much more massively built body.

In the mid-1990s, a short report mentioned a specimen of *B. cetoides* from Mississippi that contained a mass of bones and teeth inside its stomach. The remains belonged to various bony fishes and sharks, ranging up to approximately 50 centimeters long. This seemingly confirmed that *Basilosaurus* was an active predator that would hunt, catch, and consume fish in a manner apparently analogous to orcas.

Years later, in 2019, the first evidence of stomach contents in *B. isis* was revealed in a newly discovered skeleton found in 2010 at the famed Wadi Al-Hitan site, known as the "Valley of Whales." Situated in the Western Desert of Egypt, roughly ninety miles southwest of Cairo, about 38 million years ago this area was a shallow, tropical sea populated by early whales. The site is world famous for the often-complete skeletons of whales like *Basilosaurus*. In case you ever find yourself in Egypt, Wadi Al-Hitan is a protected UNESCO World Heritage Site that you can visit; many of these whales sit in the same spot where they were discovered.

Stomach contents preserved inside this new specimen showed that *Basilosaurus* not only fed upon large, 1-meter-long fishes called *Pycnodus* but, most incredibly, that it fed on closely related basilosaurid whales called *Dorudon*. These smaller whales were about a quarter of the length of *Basilosaurus*, with the species *Dorudon atrox* measuring 5 meters long. At least thirteen skeletal elements, including various bones and teeth belonging to *Dorudon*, were found in and around the stomach region of this *Basilosaurus*. Based on their size and the types of bones preserved, it indicates that a minimum of two juvenile *D. atrox* were present, each with a body length estimate of 1.5–2 meters.

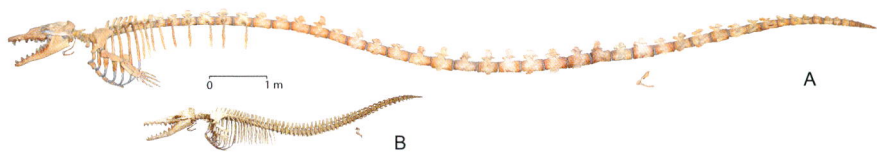

FIGURE 6.3. (A) A comparison of the fully grown adult skeletons of *Basilosaurus isis* and *Dorudon atrox*. (B) A *Basilosaurus* skeleton in the desert at the Wadi Al-Hitan site in Egypt.

([A] Image from M. Voss, et al., "Stomach Contents of the Archaeocete *Basilosaurus*: Apex Predator in Oceans of the Late Eocene," *PLOS ONE* 14 (2019): e0209021; [B] courtesy of Mohammed ali Moussa, Wikimedia Commons)

Prior to this discovery, it was presumed that *Basilosaurus* probably hunted its contemporary, not only because of its size and apex predator status, but because several subadult *Dorudon* had previously been found with bite marks consistent with predation by *Basilosaurus*. Some of the *Dorudon* remains from the stomach contents also exhibited evidence of being bitten or broken into smaller pieces, this being highly likely to have been undertaken by the *Basilosaurus* when it chowed down. The bite marks are also similar in dimension, shape, and outline to those bite marks observed in the other subadults, providing further evidence that a *Basilosaurus* also attacked those individuals. In any event, this very fossil directly confirms that *Basilosaurus* preyed upon its much smaller cousin. If you ever watched the 2001 BBC series *Walking with Beasts*, one of the episodes focused on a *Basilosaurus* that was hunting, you guessed it, *Dorudon* and its young.

The discovery is the first direct evidence of a predator-prey relationship among the two most commonly found whales at Wadi Al-Hitan, each of which is represented by hundreds of skeletons. Curiously, all known *Basilosaurus* fossils found here are adults, whereas the remains of *Dorudon* range from neonatal juveniles to adults, which has resulted in the site being interpreted as a *Dorudon* calving area. The confirmed stomach contents of at least two juvenile *Dorudon* combined with bite marks on others suggest *Basilosaurus* actively hunted in the calving area, probably targeting inexperienced youngsters and inflicting a fatal, crushing bite to the head. This is not to say that *Basilosaurus* did not attack and kill adult *Dorudon*, too, which was surely on the menu, but we simply do not have the evidence for that, at least not yet.

The fact that we have ancient apex whales from an extinct family that were almost twice as long as a bus dining on the calves of their close cousins echoes the behavior of today's apex orcas, which are well known to separate and kill young cetaceans. No sooner had whales come to dominate the ancient oceans did it become a whale-eat-whale world.

FIGURE 6.4. (Overleaf). *A Killer in the Shallows*

A monstrous toothed whale, *Basilosaurus isis*, swoops in and grabs an unfortunate juvenile *Dorudon atrox*, a much smaller species of basilosaurid toothed whale.

THE EARLY BIRD GULPS THE FRUIT

As far as we know, the first birdlike dinosaurs evolved during the middle to late Jurassic and the most iconic and arguably most scientifically important of all birdlike dinosaurs, or *avialans*, is *Archaeopteryx*. Described in 1861, it was for a long time considered to be the "earliest bird," although paleontologists still hotly debate whether *Archaeopteryx* and other early birdlike dinosaurs should be considered birds or very close relatives and cannot quite agree on which was the oldest (but let us not get into that here).

Nonetheless, paleontologists have long been fascinated by early birds and birdlike dinosaurs, especially in trying to unravel the roots of their evolutionary tree. Over the past thirty or so years, major finds in China have dramatically helped to better our understanding of their early evolution and behaviors.

In 2002, a study was published that described "a large basal bird" collected from the famous Jehol Biota of western Liaoning Province in northeastern China. Early Cretaceous in age, around 120 million years old, this area is now very famous for producing thousands of feathered dinosaurs, including birds. This small, pheasant-sized theropod was named *Jeholornis prima* and had a rather long, bony tail akin to the tails of dinosaurs like *Velociraptor* and kin (dromaeosaurids), although *Jeholornis* could fly and was presumably arboreal. This was an exciting discovery because it was another primitive species linking nonavian theropods with birds, yet a key part of the study was focused on dozens of "seeds" (ovules) associated with the specimen. An ovule is a seed's developmental precursor or immature version. For ease, we will call them seeds.

Nestled in and around the abdominal region of this *Jeholornis* are more than fifty rounded, pea-sized seeds all bearing the same shape and differing only very slightly in size, averaging 8–10 millimeters wide. A few are scattered as if they have burst out of the stomach cavity, reminiscent of the *Alien* movies, but most probably resulting from decomposition. These types of seeds were preliminarily thought to be ginkgo-like, and isolated seeds had been reported previously by paleobotanists and referred to as *Carpolithus*, but the exact group of plants they belong to is still unknown. However, magnolia and ginkgo plants have been found in the same rocks.

The preservation of undigested seeds inside the stomach provides direct evidence for seed-eating behavior in this early avialan. It may also suggest that *Jeholornis* had a crop, the part of a bird's digestive system that is used to store excess food before it enters the stomach, though some researchers think that a crop was absent in *Jeholornis*. By 2002, many hundreds of Mesozoic birds had been found, but none had yet to reveal stomach contents of seeds or any definite information about their diets. This *Jeholornis* was the first, but it was not the last, because years later three additional specimens were also found to contain seeds, including one that appeared to have ingested three different types. To date, more than a hundred *Jeholornis* fossils have been found, including some with gastroliths (stomach stones), although no specimen contains both seeds and gastroliths.

We know that many birds today eat seeds, either cracking them open with their tough beaks, grinding them up in their stomachs or eating the entire fruit with the seed inside. *Jeholornis* had only a few small, delicate vestigial teeth but was obviously capable of consuming these seeds. But how? Were these birds only eating the seeds, or were they swallowing the whole fruit with the seed inside?

Because the seeds remain completely intact inside the stomach, it is clear that it consumed them whole, rather than breaking them up. In the original study, the researchers Zhonghe Zhou and Fucheng Zhang of the Institute of Vertebrate Paleontology and Paleoanthropology in Beijing discussed the difficulties of determining whether *Jeholornis* ate whole cones on a tree, plucked the seeds from intact cones, or ate shed seeds.

To attempt to solve this mystery, in 2022, a team led by Han Hu of the University of Oxford, England, reexamined the seeds and described a new specimen of *Jeholornis* that included an exquisitely preserved skull. High-powered scans and digital reconstructions of the skull, along with direct comparisons of the mandibles of many modern birds (including types that grind and crack seeds and eat whole fruits), revealed that *Jeholornis* ate whole fruits for at least part of the year. Thus, only the harder seeds remained behind.

They surmised that *Jeholornis* ate fruits only during the seasons they were available and presumably switched to another food source at different times of the year. Further support for this interpretation may come from findings in 2023 of another *Jeholornis* with leaves from magnolia-like flowering plants and other types inside its stomach. Not only does *Jeholornis* provide the earliest evidence for fruit consumption (frugivory) in birds, but the discovery reveals that birds likely played an important role in the dispersal of seeds and fruit-bearing plants during the earliest stages of bird evolution and diversification. Even further evidence was described in late 2024 when yet another Jehol bird was reported with direct evidence of seeds inside its abdominal cavity. This time, two specimens of an unusual, toothed bird known as *Longipteryx* contained multiple rounded seeds resembling those also found in *Jeholornis*.

Birds and plants have had a coevolutionary relationship for millions of years. Both have greatly benefited from each other over time, with plants yielding nourishment for the birds and birds returning the favor by dispersing seeds in their droppings and helping plants to conquer far and wide. Eating seeds and fruits is a behavior that we find synonymous with birds today; whether they are fluttering around in the woods finding juicy berries or pecking away at birdfeeders in gardens, this relationship has been forged over millions of years of evolution and does not seem to be going extinct anytime soon.

FIGURE 6.5. (A) The complete holotype of *Jeholornis prima* with over fifty *Carpolithus*-like seeds (ovules) inside its stomach. (B) A close-up of some of the pea-sized seeds.

(Courtesy of Zhonghe Zhou)

FIGURE 6.6. (Opposite). *Seeds Are on the Menu*

The early birdlike dinosaur, *Jeholornis prima*, eats seeds among the foliage of a ginkgo tree. Below, his companion sows some seeds with a ready supply of fertilizer.

WHEN THE TABLES TURN

A predator actively hunts and kills prey to provide food for itself or its growing family, although some animals might just kill for the sake of killing and can be total jerks. As the nature of, well, nature, predator-prey relationships are everywhere throughout the animal kingdom, such as a fox hunting a rabbit or a lion stalking a zebra, but sometimes the hunter might become the hunted. No, a rabbit is not going to turn around and attack, kill, and consume the fox, but many instances exist where two predators might hunt the same prey or each other and result in one of them becoming supper.

Another way of looking at this is with those predator-prey relationships where two groups of animals each have many members that feed on the other, such as fish that eat squid and squid that eat fish. All of this might seem like the norm, and it is, but such relationships developed over many millions of years, and we have some exceptional fossil evidence capturing this type of behavior in action.

If you have ever been searching for fossils in Mesozoic-aged marine rocks, then chances are you probably found or are at least aware of bullet-shaped belemnites. This common part of the fossil is called the guard or rostrum and functioned as the internal skeleton (the hard part) for these squishy, squid-like coleoid cephalopods. They are relatives of modern-day squid, octopus, and cuttlefish. In very rare cases, belemnites and their kin are found complete with soft parts still preserved, sometimes including evidence of their arms lined with numerous sharp hooks (think of the squid stars you met earlier). One group of close belemnite relatives, called diplobelids, have also been found with outstanding preservation of soft tissues, although some fossils have a little extra: their food.

Four complete adult specimens of a diplobelid called *Clarkeiteuthis conocauda*, all collected from the roughly 180-million-year-old Early Jurassic rocks of Holzmaden in southern Germany, were each found with a bony fish called *Leptolepis bronni* in their arms. Only one of the *Clarkeiteuthis* specimens was described in detail and has a total length of 21 centimeters, almost twice the length of the trapped *Leptolepis* prey, which measures about 12 centimeters. In each example, the hooked arms of the squid-like diplobelid are very clearly wrapped around the small fish, and none of these occurrences represents a random coming together of two different species.

Living coleoids are voracious predators that quickly whip out their long, strong arms and tentacles (if they have them) to snag fish and other creatures; various species have suckers or sharp hooks that they use when hunting. Once a fish is firmly trapped in their arms, there is usually no escape. For comparison with *Clarkeiteuthis*, the living northern shortfin squid, found in the northwest Atlantic Ocean, usually captures a fish with its arms pinned around the body, leaving the head and tail protruding outward. This is a similar position observed in each of the fossils. Moreover, using its sharp, pointed beak the living squid delivers a fatal bite by cutting the spine near the head of the fish. In three of the four *Leptolepis*, there is a distinct kink in the spine near the beak of *Clarkeiteuthis*, suggesting this form of attack was probably also used to quickly kill the fish.

Unfortunately, it did not end well for either the Jurassic fish food or the squid hunters. As the research team discussed in their 2019 study, it appears that after completing a successful hunt, capturing and holding onto the fish in well-oxygenated waters, the hunters became distracted by their fish food and descended into deeper, oxygen-depleted waters, where they suffocated before coming to rest on the inhospitable dead zone of the seabed. This phenomenon is known as "distraction sinking," where a species is focused on either mating, feeding, or fighting, and has been reported in living coleoids, such as the northern shortfin squid and the giant squid.

The fact that there are at least four examples shows that this was a recurring squid-hunting-fish scenario that must have played out countless times and suggests that *Leptolepis* was a common target for *Clarkeiteuthis*. In fact, just a year after this study was published another specimen of *Clarkeiteuthis* was found associated with a fish, but this time from the slightly earlier Jurassic rocks near Lyme Regis, in Dorset, UK. This *Clarkeiteuthis* is a separate species and contains a different, albeit similar-sized small fish trapped in its arms. This additional example provides further evidence for such a predator-prey relationship.

Remaining underwater in Holzmaden's Early Jurassic sea, another similar case, but with a twist, was the discovery of a belemnite that was found with parts of a decapod crustacean in its arms. This research was published

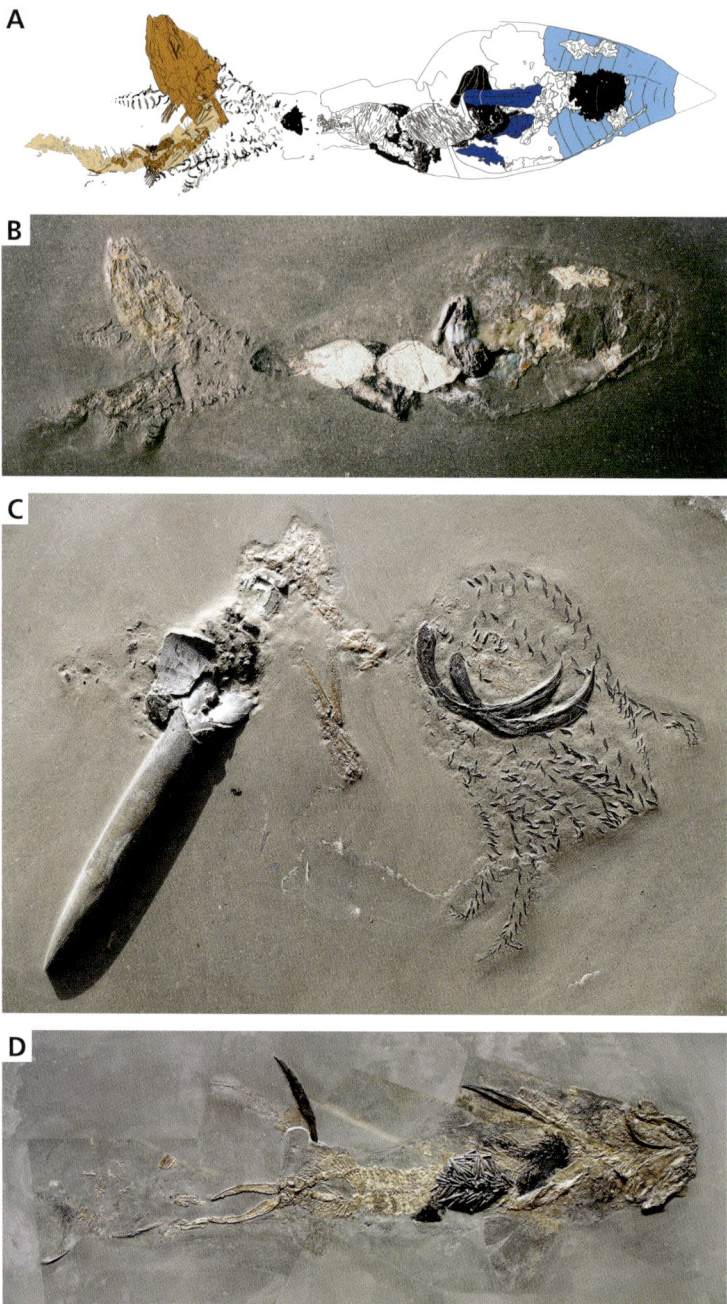

FIGURE 6.7. (A) Interpretive illustration and (B) photograph of a *Clarkeiteuthis conocauda* with a *Leptolepis bronni* caught in its arms. (C) The crunched belemnite *Passaloteuthis laevigata* with its last crustacean (*Proeryon*) meal in its arms. (D) The exceptional skeleton of the shark *Hybodus hauffianus* with a belly full of about two hundred belemnites. All are from the Early Jurassic of Holzmaden, Germany.

in 2021 and led by the cephalopod expert Christian Klug from the University of Zurich, who was also part of the team in the aforementioned study. The belemnite belongs to the common species *Passaloteuthis laevigata* and is almost complete, including more than four hundred microhooks on its arms and two strongly curved megahooks, called onychites. The crustacean is a type called *Proeryon* and represents either a fresh catch or, perhaps more likely, a cast-off molt that is trapped in the arms of the squid, with parts of its claws easily distinguished. The squid had begun tucking into and taking apart the decapod but did not manage to finish its last supper. The problem? There's always a bigger fish.

In this situation, the predator became prey. Though the belemnite is almost complete, it preserves only some of the soft parts, and the posterior end of the rostrum is damaged resulting from a direct attack by a much larger predator. Although identifying the culprit is challenging, the team surmised that it was likely a shark. For this unlucky belemnite, the predator landed a near-perfect bite, enough at least to tear off much of the squishy bits. In return, we have "leftover falls," the remains of food that a predator left behind for an unknown reason. This renders the belemnite a piece of unfinished food. Such food falls are known among animals in today's oceans, such as the tail or fin of a fish being left or missed while the rest is eaten by the predator. For these kinds of fossils, the team defined a new technical term, "pabulite," meaning food stone.

Direct evidence for the predation of belemnites by a shark comes from a phenomenal fossil from Holzmaden of a type of shark called *Hybodus hauffianus*, which has a huge mass of an estimated 200 (93 visible) belemnite rostra inside its very full belly. I have examined this specimen firsthand on display at the Natural History Museum of Stuttgart, and it is truly beautiful. Thinking about this shark consuming about two hundred complete animals; that is a staggering number of squid to eat without even chewing off or regurgitating the hard parts. Did this greedy shark die from

FIGURE 6.8. (Overleaf). *The Hunter Becomes the Meal*

A shark (*Hybodus hauffianus*) bites into a belemnite (*Passaloteuthis laevigata*), which is holding and eating a crustacean (*Proeryon*). A food chain in action.

overconsumption? Almost certainly. The accumulation of the indigestible hard rostra has been interpreted as the likely cause of its death, so the squid had the eternal last laugh.

It can be presumed that prehistoric fish would have eaten squid and vice versa, but here we have direct evidence for this behavior in exactly the same formation and time frame. The occurrence of five specimens of the Jurassic squid-like *Clarkeiteuthis* with its prey represents the oldest-known evidence of cephalopods preserved with their vertebrate prey. On the flip side of the dinner table, the probable shark that attacked the squid eating a crustacean provides a snapshot of a fossil food chain in action, showcasing the partially leftover predator turned prey preserved with its excess food. These are exceptional fossils, and one thing is for sure. The *Hybodus*, with a belly full of belemnites, probably regretted its decision to overindulge. Or maybe not.

7 | Conflict

Dramatic footage captured a female snow leopard patiently stalking a group of bharal, or "blue sheep," along a snowy cliffside in the Himalayas. Suddenly, boom! The chase is on. Just as the leopard captures its targeted prey, the pair fall over the edge, plummeting some 300 feet. Crashing and rolling against rocks and snow as they drop, tussling and turning in midair, the cat does not let go. Eventually, the pair land, and the cat somehow survives the ordeal and begins to tuck into its prey.

TACO TAKEDOWN AND THE ETERNAL "HUG"

One of the most extraordinary dinosaur discoveries ever made was uncovered in 1971 as part of an expedition to the Gobi Desert, in Mongolia. This legendary fossil captures a spectacular showdown of a *Velociraptor* and *Protoceratops*, battling to the very end. This fossil was the inspiration for the cover of *Locked in Time*. When deciding on *the* fossil to really symbolize the book, I could not look past the "Fighting Dinosaurs" simply because of the story it captures—a literal pair of fighting dinosaurs—and the truly unique nature of the fossil.

However, as you may also be aware, there is another apparent dino fight of what is thought to be an adolescent *Tyrannosaurus rex* and *Triceratops* that might have been battling it out. This aptly named "Dueling Dinosaurs" fossil was discovered in Montana, but its secrets have yet to be revealed (although it should be soon). So, technically, we have one definite dino fight and one that is possibly a fight caught in time. But now, wait for it . . . we have the mammal that took down a dinosaur.

In the early 2000s, a cat-to-badger-sized mammal from the Cretaceous was described, called *Repenomamus*, which represented one of the largest mammals from the age of dinosaurs. Well, in what is easily one of the most extraordinary fossil finds in recent times, even more unexpected than those mentioned above, one *Repenomamus* was found in direct conflict with a complete example of the Labrador-sized ceratopsian dinosaur *Psittacosaurus*, the same type of dinosaur (or "tacosaur") that we met in chapter 1.

On July 18, 2023, I opened Twitter for the first time in a while, and on my timeline, staring at me, was this epic fossil. Fortuitously, the research had literally just been announced publicly, and my timing could not have been any better, as I was in the midst of writing this book and deciding on the final fossils for this chapter. Jordan Mallon, a paleontologist at the Canadian Museum of Nature in Ottawa, Ontario, and a member of the research team, had just shared a photo of the fossil along with these words: "Some exciting news to share this morning."

I was blown away. So much so that I just had to clear part of my afternoon to discover more and drool over this fossil. Subconsciously, I had already decided to include it in this book and, unable to contain my excitement, began writing about it just two days later. But what exactly is happening in this fossil to make it so special? Does it actually represent a fight to the end, or perhaps it is just an act of scavenging or simply a chance occurrence?

First, a few basics: The fossil was discovered on May 16, 2012, near Lujiatun Village in Liaoning Province, China, an area famed for its fossils, including many *Psittacosaurus* specimens. Collected from 125-million-year-old Early Cretaceous rocks, it was subsequently acquired by the lead author of the new study, Gang Han, who donated it to the Weihai Ziguang Shi Yan School Museum in 2020.

When you first gaze upon this fossil you are immediately drawn to just how complete it is—there are *two* intertwined skeletons. The *Psittacosaurus* is lying on its front, head turned to the left, hind limbs folded by its body, and tail curled around and pointing toward the head. From the tip of its parrot-like beak to its tiny toes, every single bone is present. The body of the mammal is coiled to the right and spread across part of the dino's body, toward the head and over the left arm; only the end of the tail is missing. Both are so complete that the species can be positively identified as *Repenomamus robustus* and *Psittacosaurus lujiatunensis*.

There are three key things happening in the fossil that can instantly rule this out from being a sheer coincidence, such as, for example, that they were transported and washed together. Notably, the left hand of the *Repenomamus* is in the mouth of the *Psittacosaurus*, specifically with its fingers clutching the lower jaw. The mammal's jaws are clamped around two of the dinosaur's ribs, with its teeth clearly embedded. The left foot of *Repenomamus* is also trapped in the folded left leg of the *Psittacosaurus*. Then there is the dominant position of the mammal atop the dinosaur as if using its position to subdue it. There is no denying that this shows direct interaction between the pair, with the mammal clearly the aggressor, but reliably interpreting whether the pair are locked in mortal combat or if the mammal was scavenging is hard to assess.

The team cited three lines of evidence that favored predation over scavenging. First, there is not a single bite mark present on the dinosaur, with such marks typically being left behind by scavenging carnivores. Next, they cited the unlikelihood that the two individuals would become so intimately positioned if the dinosaur was dead prior to *Repenomamus* appearing on the scene. Last, and sort of connected to the first point, if the mammal was scavenging, then why would it be present atop the body of the dinosaur when it could easily have tucked into its dinosaur meal from ground level? Fair points.

Further to all the interpretations and observations, the CSI-like scenario could point to a combination of these things. For instance, you could question if this really was a pair dying mid-fight, then how or why did the mammal's hand not get severed or sliced by the tough beak of the *Psittacosaurus*? Surely, placing your hand in the beak mid-fight would have been a dangerous place. The same goes for biting the ribs: How were those ribs exposed? Did the mammal bite from the side and up and hang on? Such observations could infer that perhaps rather than scavenging, the mammal had indeed killed the dinosaur, only to begin eating it and perching itself above the body, hence the placement of the hand in the mouth, then tucking its head in low with the jaws around the ribs. Or maybe it was trying to free itself when the dinosaur collapsed and trapped its foot. The team even considered whether the *Repenomamus* may have been eating the *Psittacosaurus* while it was still alive, tucking into its live prey in a similar vein to what some modern species do, such as African wild dogs or spotted hyenas.

Perhaps another way of looking at this is the great difference in size between the two. The *Psittacosaurus* is much larger than the *Repenomamus*, with a body length of just over a meter (119.6 centimeters), and the mammal measures almost half a meter (46.8 centimeters), so a difference of more than 50 percent. As for weight, estimates put the former at more than 10 kilograms and the latter at about 3.5 kilograms; thus, the dinosaur is estimated to be more than three times heavier than the mammal. Based on comparisons with many other *Psittacosaurus* fossils, the estimated age of this individual was 6.5–10 years old, with the latter being more likely, and thus representing at least a subadult. Repeno was probably also a subadult when it died. Focusing on these differences, you might think that the dinosaur was simply too large for the mammal to take on and, therefore, must have been scavenging a carcass. Though, hold your horses.

The team wanted to address whether the dinosaur was too large to have been a realistic target for the mammal, so they examined predator-prey relationships (and extreme bodyweight comparisons) in terrestrial carnivores that could be used as good analogues. They looked at lone wolverines (not the Hugh Jackman kind) that are known to occasionally hunt animals much larger than themselves, such as domestic sheep, reindeer, and even moose. The common weasel was another mini predator referenced, which is known to attack much larger animals such as grouse and hare. Mustelids in general are known for taking on prey that may be far larger, even more than ten times their size. So, the size discrepancies should not be a deciding factor in determining the interaction. Combining this with their observations, the team favored predation over scavenging.

Poor *Psittacosaurus*, "won't somebody please think of the children?" Well, another *Repenomamus* did think of the children and ate one of them, even dismembering it for good measure. This is based on a unique fossil described in 2005 that showed this individual had eaten a baby *Psittacosaurus*, whose partially digested remains were found inside its gut. This confirms that *Repenomamus* already had a taste for tiny tacosaurs, but this latest fossil shows that it challenged much older individuals, something that I had even said was unlikely given the difference in size. Yes, I clearly underestimated this mighty mammal, and I am happy to be wrong, especially when it leads to fossils like this. We could speculate for days as to *why* this interaction happened, such that perhaps the dinosaur was protecting a nest or something similar, but we will never know the reasons why this interaction came to be.

Speaking of which, how exactly did this fossil come to be so exceptionally preserved with such lifelike detail? The preservation of many fossils found in the same area and from the same geological formation has long been linked to volcanic activity, including ashfalls, volcanic mudflows (lahars), and pyroclastic flows. Since the description of this fossil, very recent work has argued that many of these fossils were instead the result of burrows suddenly collapsing, thus entombing the inhabitants. Although, this interpretation has already been challenged.

Nevertheless, the research team proposed that the mammal's attack may have been caught in the act by a sudden lahar-type volcanic debris flow that smothered and entombed the pair, which would have rapidly preserved them in a sort of freeze-frame, just as they are. This type of situation would help to explain how the skeletons became entwined, with the left hindfoot

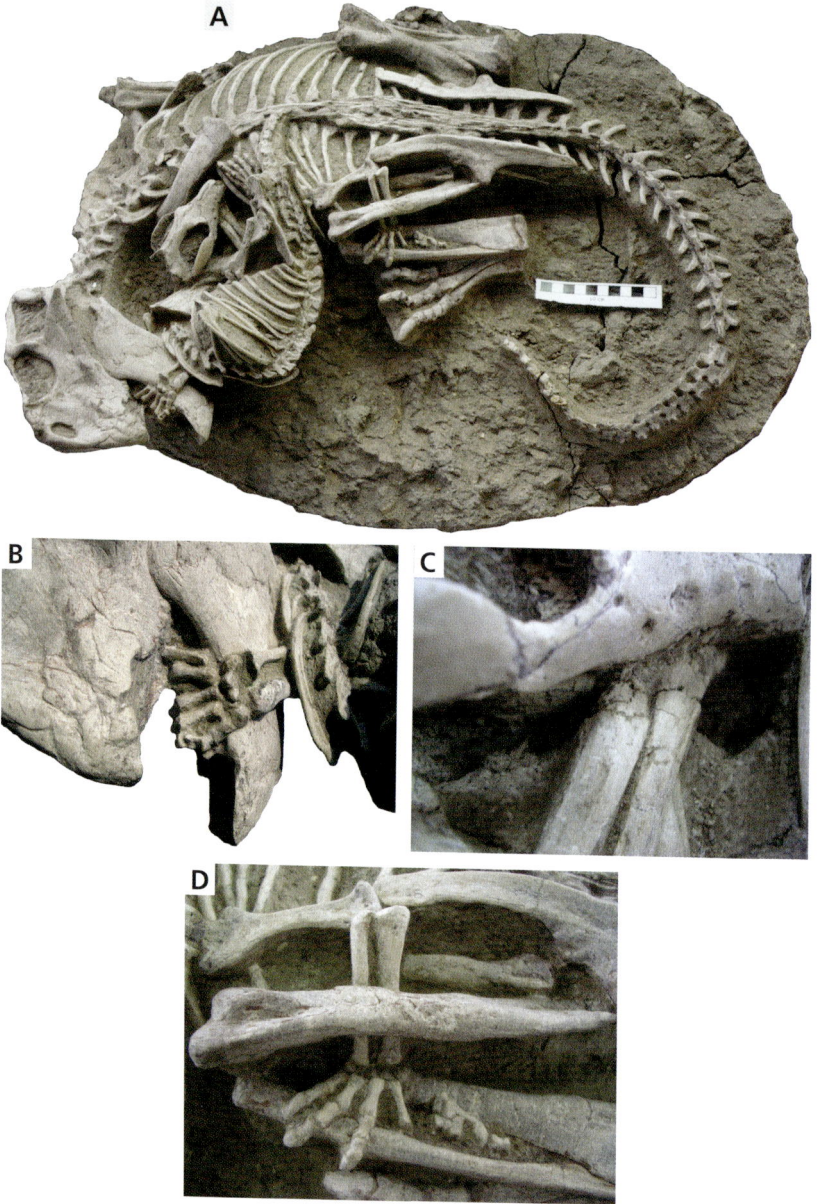

FIGURE 7.1. Their final bout. (A) The incredibly preserved "fighting fossil" of a mammal, *Repenomamus robustus*, and a dinosaur, *Psittacosaurus lujiatunensis*. (B) The left hand of the mammal can be seen grabbing the dinosaur's jaw. (C) The mammal's jaws are between two of the dinosaur's ribs and the teeth are clearly embedded. (D) The mammal's left foot is trapped between the folded left leg of the dinosaur.

(Courtesy of Jordan Mallon)

of *Repenomamus* trapped under the folded left leg of the *Psittacosaurus* when it collapsed to the ground. Plus, there are no tooth marks or other scavenging signs on either skeleton.

During the reign of the dinosaurs, mammals pretty much took a back seat. They remained small, usually mouse-sized, critters and are often portrayed as fleeing for their lives because hungry dinosaurs were waiting around every corner. The latter was not really the case, but it might be what you read in many earlier books or observed in documentaries. I can imagine the number of celebratory fist pumps that many mammal paleontologists did when the news of this discovery was announced (if you are reading this, I see you).

It is interesting to think that we now have three fighting (or at least in direct association) dinosaur examples, one here with a mammal, but all three including a ceratopsian. Clearly, the ceratopsians could not catch a break and often found themselves in the wrong place at the wrong time. In the case of this fossil, and considering that we already have the evidence of *Repenomamus* dining on a baby, perhaps *Psittacosaurus* were commonly targeted by these early predatory mammals. This fossil convincingly shows that these mammals could pose a genuine threat. It is a scenario to ponder some more, but it does change our understanding of dinosaur-mammal dynamics during this particular slice of prehistoric time.

This is truly a special discovery that captured the public's imagination, the sort of fossil that, as a paleontologist, I would very much have enjoyed studying and describing. I cannot help but feel a little jealous of the team that worked on it. I do, however, want to add a side note to this story, which surrounds many of the comments on social media. Most expressed their amazement at such a remarkable fossil, but alas, many were quick to say "nope, that's wrong," or cry "fake". To a point, the latter might be understandable because many fossil fakes have come out of China and often look too good to be true.

FIGURE 7.2. (Overleaf). *A Mid-Fight Flight*

A ravening Cretaceous mammal, *Repenomamus robustus*, has taken on a much larger dinosaur, *Psittacosaurus lujiatunensis*. When the *Psittacosaurus* attempts to shake off the attacker, the *Repenomamus* holds on with its teeth!

However, the team spent months, perhaps years, analyzing this fossil to come to the conclusions presented. To quote a certain movie, "So you know, try to show a little respect." Plus, the lead author, Gang Han, has strong connections with the locals in the Lujiatun area. He also happens to know the person who prepared the fossil, who is a retired technician from the Institute of Vertebrate Paleontology and Paleoanthropology (IVPP) at the Chinese Academy of Sciences, one of the world's leading paleontology institutions.

Sometimes, rather than jumping straight in and crying "fake," I would like to think those people might sit back, smile, and think, "Wow, this is an exceptional discovery from a time when our planet—*our home*—was occupied by these marvelous creatures."

FOUR DAGGERS BETTER THAN TWO

Think of Ice Age megafauna, and *Smilodon*, mammoths, mastodons, and maybe giant sloths immediately come to mind, and not just because of the *Ice Age* movies. *Smilodon* is one of the most iconic prehistoric predators of all time. About the size of a tiger, but with a bearlike muscular build, it famously had huge, dagger-like, curved canine teeth that were more than 25 centimeters in length. Even the generic name is on point, as *Smilodon* loosely means "knife tooth."

While people may often refer to *Smilodon* as the "saber-toothed tiger," these animals are *not* tigers. On the great cat tree of life, otherwise known as the Felidae family, which includes all of the approximately forty living species of cat and all extinct members, *Smilodon* does not sit closely to the tiger subfamily part of the tree known as Pantherinae. Rather, *Smilodon* belongs to an extinct subfamily of cats called Machairodontinae. Think of it this way: All tigers and kin are cats but not all cats are tigers. Plus, *Smilodon* is just one of many saber-toothed cats because there are several species known, mainly in the Machairodontinae, which had elongated dagger-like teeth.

Smilodon proper, the OG of the saber-toothed cat world, lived in various locations across North and South America from 2.5 million to 10,000 years ago and is known from three different species. In the popular realms, when we think of *Smilodon* the generic image that we conjure very likely comes from or is inspired by the famed species *Smilodon fatalis*, known from several thousand individuals collected at the celebrated La Brea Tar Pits in California. As of writing this, I have just returned from a trip to Los Angeles and a visit to the tar pits and museum. There is definitely a buzz around *Smilodon*, having overheard multiple times, "Where's the saber-tooth?" or "Where's that thing with the big, pointy teeth?" *Smilodon* is also the state fossil of California.

Based on the fantastic collection of *Smilodon* remains found in the sticky tar pits, we have learned a lot about this ancient kitty's anatomy and various bits about its behavior, such as that it might have lived in groups and maybe hunted in packs. One of the perhaps more sinister aspects of its behavior, however, comes from the largest and first-to-be-named species,

Smilodon populator, a South American native originally described in 1842. The species name, *populator*, translates as "the destroyer"—for good reason.

This felid's most distinctive feature, the bladelike, hypertrophied canines, have been the focus of intense study for a very long time and still remain under debate. The teeth were no doubt used as deadly weapons, but for as long as we have been fascinated by them, scientists have argued about how exactly they were used and how powerful they really were. Did they deliver a deep, penetrative bite or a targeted stab to the throat or belly, or were they used to slash at prey? Maybe it was a combination of these things and more. In a study published in 2019, led by Nicolás Chimento of the Bernardino Rivadavia Museum of Natural Sciences in Argentina, his research team presented evidence of combative interactions between two new specimens of *S. populator*, one collected from Buenos Aires Province and the other from Córdoba Province, Argentina.

The evidence? This pair of *Smilodon* fossils have traumatic skull injuries that appear to have been inflicted by the canines of another *Smilodon*! Both specimens bear a distinct, elliptical-shaped opening in the front half of the skull, between the eyes, which measures 3 centimeters long by a maximum width of 1.5 centimeters. In both specimens, the posterior end of the hole is offset toward the left side of the skull and a depressed area of bone surrounds the holes. Matching the holes with *Smilodon* teeth, the size and general contours of both the injuries and the canine teeth are consistent, so much so that when a *Smilodon* canine is placed into the hole, it matches perfectly in shape and size. There is a somewhat strange Cinderella slipper analogy here.

The specimen from Córdoba has minor, depressed fractures around the main hole and shows no signs of regrowth, suggesting this animal probably died as a direct result of the bite. The Buenos Aires specimen shows some signs of bone healing, indicating that the individual probably lived for quite a long period following the confrontation.

Support for this *Smilodon* versus *Smilodon* interpretation comes from earlier studies that reported evidence of wounds in the skulls of other machairodontines thought to have been made by saber-toothed cats. In the 1980s, similar openings were also reported in specimens of *S. fatalis* from La Brea. A similar example compared with those discussed herein was

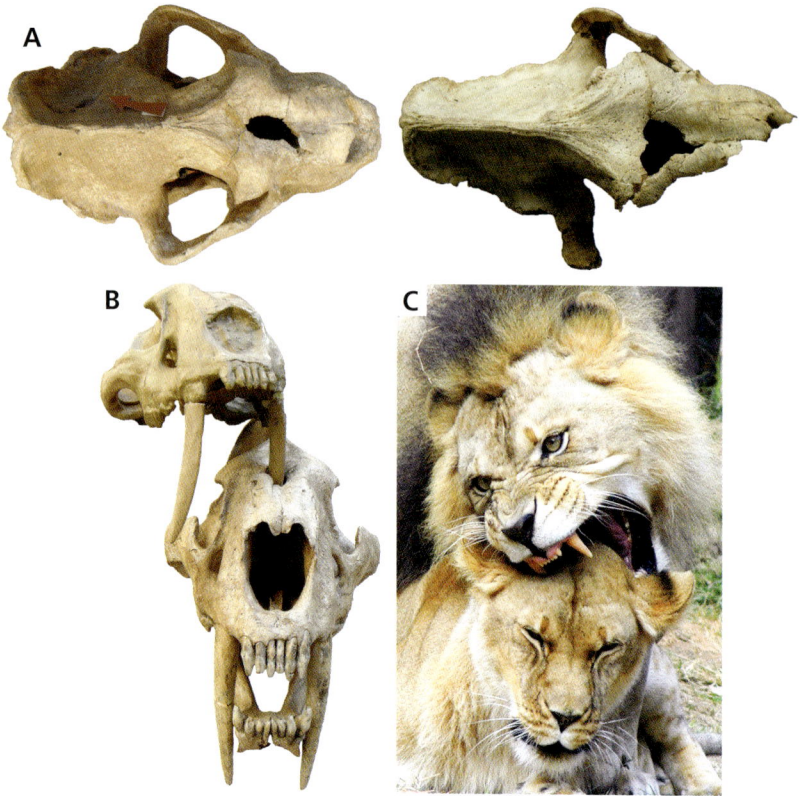

FIGURE 7.3. (A) Two skulls of *Smilodon populator* showing traumatic skull injuries. The left skull is from Buenos Aires Province, Argentina, and the skull on the right is from Córdoba Province, Argentina. (B) The same skull from Luján River with the canine of another *Smilodon* inserted through the opened injury. (C) An angry lion bites down on the top of the head of a lioness at North Carolina Zoo.

([A–B] Courtesy of Nicolás Chimento; [C] courtesy of Valerie Abbott)

FIGURE 7.4. (Opposite). *The Fatal Bite*

The battle for dominance between this pair of *Smilodon populator* ends abruptly when a dagger-like tooth penetrates the skull too deep and stabs the brain.

B.OB NICHOLLS ART

found in a specimen from Turkey of the saber-toothed cat *Machairodus*, where a hole was identified between the eyes that caused the animal's death. Once again, the size and shape of this hole were consistent with the size and shape of the dagger-like canines of *Machairodus*. Outside of intraspecific combat, a skull of a giant armadillo called a glyptodont was found with two subparallel, elliptical-shaped openings matching the size and contour of *Smilodon* canines. We can infer that they were likely to have been produced by a contemporaneous saber-toothed cat.

As the team discussed, they cannot rule out entirely that the injuries were caused by a potential kick to the skull—lions today might receive a well-timed, potentially fatal kick by their intended prey (such as a zebra)—but a single hole combined with the size, shape, and general structure of the injury suggests this is highly unlikely. The shape also does not match what we would expect for bears, canids, or other carnivores that lived at the same time, as they had more conical-shaped canines that would presumably have left a differently shaped hole.

If we look at living felids, we know that they may engage in intraspecific competition, be that fighting for dominance, territory, access to mates, food, and more. It is very probable that *Smilodon* engaged in such direct conflicts, too. Being excellent grapplers, they would have used their powerful muscles to wrestle each other to the ground and perhaps deliver a bite.

Along these lines of conflict, similar injuries to those discussed herein are common in living felids and have been reported in small-sized ocelots to cougars, cheetahs, and jaguars. Such injuries may be the result of agonistic interactions between the same sex or between males and females and may often result in death. It is probably also worth saying that male cats can be sexually aggressive, too, and bite down on the neck of the female to hold her in place when mating. Although highly unlikely to be the cause of our *Smilodon* scenario, it is interesting to ponder if *Smilodon* engaged in similar sexual biting behavior and how the canines were used or got in the way.

Additional support for this catfight theory comes from some *Smilodon* lookalikes, at least as far as having large canine teeth and catlike bodies go. This group of sleek, catlike critters is called nimravids, often nicknamed "false saber-toothed cats," considering their outwardly similar appearance, although these felid imposters are distant cat cousins. To be fair, they are not really imposters considering that they evolved the saber-toothed look several million years before the true saber-toothed cats even appeared.

Several nimravid fossils collected from ~30-million-year-old (Oligocene) rocks in South Dakota and Nebraska have been found with tooth marks across the skulls and around the eyes, like the *Smilodon* skulls. The first report of such an example goes back to the 1930s and to a fossil identified as *Nimravus brachyops*, which displayed a puncture wound in the front left of the skull, just above the eye. The bite also showed signs of healing. More recently, in 2013, Clint Boyd, the senior paleontologist working for the North Dakota Geological Survey, and a team of researchers presented a talk at the annual conference of the Geological Society of America. Their research focused on another example, this time a different nimravid, *Hoplophoneus*, where a series of puncturing bites were identified on the skull.

This specimen led Boyd and his team to ponder whether they could find any additional evidence and decided to look at nimravid skulls in museum collections and prepare and clean others. The result? They found more seemingly bitten nimravid skulls, including fossils of *Nimravus*, *Hoplophoneus*, and another one called *Dinictis*. The team also reexamined the original specimen described in the 1930s, including undertaking further cleaning, and found that the rehealed puncture mark also displays a bone infection. However, they found a second set of bite marks with no evidence of regrowth. Based on the positions of the bite marks in and around the orbits, they concluded that most attacks came from behind with clear attempts to target the eyes and, in theory, blind the opponent. Nasty stuff.

While talking about these nimravids, I must mention a fantastic fossil known as "The Innocent Assassins." It was first described in 1932 and includes an *N. brachyops* skull with the right canine tooth piercing through the humerus of another *Nimravus*! Though it might at first appear like a pair fighting to the death, or maybe an accidental happening between two scrapping nimravids, recent interpretations suggest that it could have happened postmortem when the skull and humerus were forced together during fossilization. In either case, this is a fabulous fossil.

As the apex predator of its day, confrontations with fellow combatants must have happened regularly in the life of *Smilodon*. In most cases, perhaps all that was hurt was the loser's pride, while occasional flesh wounds or more serious damage surely occurred. Although the debate rages on about how *Smilodon* deployed its famed gnashers, there is a difference between fighting to catch and eat your prey or fighting all out with a competitor. These rare fossils from Argentina, combined with previous reports of similar damage in *Smilodon* and other saber-toothed cats and false saber-toothed cats, show that these fierce felids would occasionally use their toothy daggers to stab the skulls of their rivals, sometimes with the deadliest consequences.

CROC DINO DINNER

For many people, crocodiles are the ultimate symbol of an apex predator. Watching a croc patiently bide its time, waiting for that crucial moment to unleash its powerful, bone-crunching jaws, clamping them around its prey, and immediately beginning to rip chunks out of its meal is a sight to behold. That is, witnessed from a great distance and usually from the comfort of your couch watching a nature documentary.

When I say crocodiles, I am referring to the broader group of crocodilians (or crocodylians) whose evolutionary history is nestled back more than 95 million years in the Cretaceous. But crocs and kin are part of a more comprehensive, much more archaic group referred to as crocodyliforms ("crocs" herein), which includes true crocodilians and an array of wondrous croc-like creatures whose earliest members appear much deeper in time during the Late Triassic. Despite such a long, successful history there are only around twenty-five crocodilians today, which is a far cry from their rich and diverse past, where some species were fully marine and others even herbivorous. There are more than ten thousand living species of reptiles today, with crocs making up only about 0.25 percent. Still, crocs are one of the most successful hypercarnivores of all and continue to fill key roles in ecosystems today.

If you have ever encountered a large crocodile in the wild or at a zoo, no doubt its size and powerful, toothy jaws captured your imagination. Ignoring potentially extra-lengthy anacondas and pythons, the largest living reptile is the saltwater crocodile, which can reach a massive 6–7 meters and weigh a little over a ton, although such a size is rare. Salties are the most powerful reptiles on the planet, a title that did not quite hold up for ancient crocs during the reign of the dinosaurs despite some species doubling the average size of a saltwater croc.

Several species of prehistoric croc may lay claim to the title of ultimate croc, that is, the largest of them all. A few names to throw into the ring include the enormous Miocene caiman *Purussaurus*, the long-snouted Cretaceous *Sarcosuchus*, and the alligatoroid *Deinosuchus*. Easily, these ancient crocs reached 10–12 meters and were genuinely enormous, representing the apex predators in their respective environments. Sticking with the latter two, because they lived during the dinosaur heyday, though at different times in the Cretaceous and in different places, it is curious to ponder what interactions occurred between these ancient crocs and dinosaurs. Despite leading fundamentally different lifestyles, at least in large part, they would definitely have crossed paths on occasion, with some dinosaurs perhaps playing a role as mammals do today when crossing crocodile-infested rivers and lakes.

Let us take a slightly closer look at *Deinosuchus*, one of the oldest true crocodilians and a giant relative of the modern alligator. A few years ago, I spent quite a bit of time diving deep into the literature on this giant when I was writing a children's pop-up book, *Prehistoric Beasts*, which included a section linking modern-day gators with their extinct relatives. Of course, I had previously read about *Deinosuchus*, even examining some of its bones over a decade ago, but one of the most interesting areas that caught my attention like a gator snatching its prey was the potential evidence that big Deino dined on dinosaurs.

Hundreds of *Deinosuchus* fossils have been documented from at least ten states across the United States and in Mexico, all from the Late Cretaceous and 75–82 million years ago. A lot of what we have learned about *Deinosuchus* has occurred over the past two decades or so and includes one influential book published in 2002, *Deinosuchus: King of the Crocodylians*, written by David Schwimmer. No, not that David Schwimmer (Dr. Ross Geller from *Friends*) but an actual paleontologist of the same name. Sorry, Dr. Schwimmer, if you are reading this because no doubt you have had that comparison made thousands of times.

Anyway, Schwimmer's research on *Deinosuchus* has helped to illuminate many more details about this giant ambush predator, including describing a handful of dinosaur bones, comprising hadrosaur vertebrae from Texas and part of a theropod limb bone from New Jersey, that preserve bite marks matching those of a giant gator. Based on the size, shape, and cross-section of the theropod bone, it is thought to represent a metatarsal from a juvenile tyrannosaur. This bone was so pulverized and mashed up that Schwimmer stated in his book that the bone "resembled a dog's worn chew toy"! The combination of croc tooth–shaped bite marks and their geologic age and location point to *Deinosuchus* having munched on these dinosaurs. A bit of food for thought: One study showed that *Deinosuchus* was capable of the infamous croc death roll, where it would have spun around in the water while firmly gripping its prey and twisting off chunks of meat. The unlucky animal was probably obliterated in a matter of minutes.

Though dinosaurs were on the *Deinosuchus* menu, there is more evidence that it dined largely on sea turtles, with several turtle carapaces found with bite marks where the giant gator chomped with its massive teeth. Further evidence for this comes from the teeth of *Deinosuchus* that were specifically for crushing and are often found to be highly worn and blunt, presumably from abrasion with the hard shells. Living in the extensive wetlands that bordered the coasts, it no doubt ate fish and other marine creatures, too.

We can confidently assume that *Deinosuchus* and many other ancient crocs ate dinosaurs, through active hunting but also scavenging. However, that critical, clinching piece of conclusive evidence of a dinosaur inside the gut of one of these species remained rather elusive to paleontologists, perhaps due in part to crocs having extremely corrosive stomach acids that may quickly have removed any signs of a last meal. That was, however, until recently when direct proof came with the discovery of a new species of Cretaceous croc from deep down under in Queensland, Australia. Crikey, it would inevitably be Australia, a country somewhat synonymous with crocodiles and indeed one of the places that big saltwater crocs call home.

Described and named *Confractosuchus sauroktonos* in 2022, this croc lived a few million years earlier than *Deinosuchus*, 92.5–104 million years ago, but still in the early Late Cretaceous. Collected in Winton, Central West Queensland, from one of the sites of the Australian Age of Dinosaurs Museum, the fossil comprises a skull and partial skeleton that is estimated to have had a maximum body length of 2.5 meters.

The croc is affectionately known to the research team, led by the paleontologist Matt White, as "Chooky"—Australian slang for chicken—after his team initially thought they might have found a chicken-sized theropod. However, Chooky's scientific name roughly translates to the "broken crocodile dinosaur killer," which is quite appropriate, and not just for the shattered concretion the fossil was serendipitously found in. Nestled inside the preserved gut of this *Confractosuchus* are parts of a juvenile ornithopod

FIGURE 7.5. The giant alligatoroid, *Deinosuchus*, from Utah, on display at the Natural History Museum of Utah, Salt Lake City.

(Photograph by the author)

dinosaur, including several vertebrae and parts of both femora, among other bits. One of the dinosaur's femurs has a distinctive circular tooth puncture mark matching the teeth of *Confractosuchus*, whereas the other was sheared in half. Given that the bones have been partially digested, the exact identity of the unfortunate ornithopod remains a mystery for now, but it is especially noteworthy that these are oddly the first skeletal remains of an ornithopod known from the rock formation where the croc was collected (called the Winton Formation). Chances are, this might even be a new species of dinosaur.

This croc meal revealed direct evidence of oral processing, where the croc chomped this dinosaur and then dismembered it into more bite-sized pieces (carcass reduction). As the researchers stated in their study, these features are all diagnostic hallmarks of some modern croc-feeding

FIGURE 7.6. (Opposite). *The Dinosaur Devourer*

A hunting tyrannosaur has ventured too close to the water's edge, and it pays the ultimate price. The first bite from the colossal crocodilian *Deinosuchus riograndensis* crushes the dinosaur's skull to pieces.

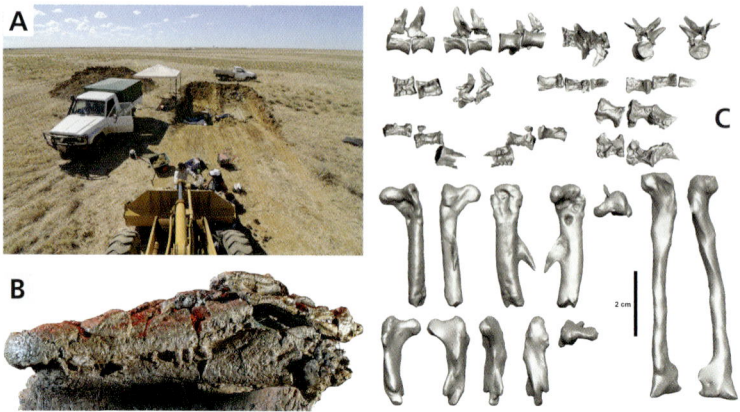

FIGURE 7.7. (A) The *Confractosuchus sauroktonos* excavation at one of the Australian Age of Dinosaurs localities near Winton, Queensland, Australia. (B) Skull of *C. sauroktonos*. (C) Digital renderings of the ornithopod dinosaur bones found in the stomach of *C. sauroktonos*.

(Courtesy of Matt White)

behaviors. Based on an analysis of the skull, the team found that although a dinosaur was found as the croc's last meal, this would not have formed its typical diet, and it was thus unlikely to be a dinosaur-eating specialist. Rather, *Confractosuchus* was more of an opportunistic generalist capable of taking down prey larger than itself, akin to several modern crocs. Given the young age of the ornithopod, it might be assumed that it veered too far away from a herd and into the deadly territory of this croc, or perhaps *Confractosuchus* targeted young, inexperienced juveniles when they were passing through the area.

Crocs eating their cousins, the dinosaurs, is a behavior that has continued through the ages and into the present day. Evidence shows that *Deinosuchus*, one of the largest crocs of all time, if not *the* largest, took down big-game dinosaurs, from large herbivores to even tyrannosaurs that may wandered too close for comfort. Even if dinosaurs were not part of the usual diet, *Confractosuchus* undoubtedly confirms that they were on the menu from time to time.

SMASH THAT

Big teeth, long claws, whippy tails, spear-like spikes, and bone-crunching bites are just some of the weapons in the dinosaur arsenal. But what is that one chunky, bumpy, and iconic thing missing that might be at the top of your list if tasked with picking the coolest piece of dinosaur kit? Yup, the massive bony tail club, a legendary weapon from that magnificent group of quadrupedal, armored dinosaurs, the ankylosaurs.

With their low-slung, robust, and wide bodies covered with an extensive coat of armor from head to tail, and with some but not all members rocking a tail club, it is safe to say that the ankylosaur family was the most heavily armored of any dinosaur group, even more so than their spiky stegosaur cousins. The North American, Late Cretaceous *Ankylosaurus* is the poster child of these tanklike dinosaurs, a quickfire name said by dinosaur fans of any age, which is also the namesake for the wider ankylosaur group. Despite being a renowned feature of ankylosaurs, the tail club appears to have evolved relatively late in ankylosaur evolution, as exemplified by *Ankylosaurus*.

The tail club is formed of stiff, tightly interlocking vertebrae that together create a "handle," and it terminates with large, bulbous osteoderms that form the prominent club or knob that enveloped the end of the tail. Besides ankylosaurs, tail club–like structures are rare among vertebrates. They have been found in a couple of sauropods with strange, small tail clubs, one extinct family of turtles called meiolaniids, and famously in some car-sized armadillos called glyptodonts.

Having a whopping big, clublike bony mass at the end of the tail is a pretty strong indicator that this structure would have been used as a weapon. Makes sense, since in the case of *Ankylosaurus*, it shared its ecosystem with none other than *Tyrannosaurus rex*. Having rows of thick bony armor and a rock-hard tail club are two solid ways to combat an attack from a *T. rex*. There is a bit of a trope in paleo art that depicts an *Ankylosaurus* waving its tail club in the direction of a *T. rex*, which absolutely must have happened, and is something we all did with toys as children (and my office shelf with both toys confirms that I still do this today).

It has long been assumed that ankylosaur tail clubs must have been used as defensive weapons against attacking predators, like a *T. rex*, although it was not until 2009 that the ankylosaur aficionado and authority Victoria Arbour, now the curator of paleontology at the Royal BC Museum in Canada, decided to tackle this question head-on. Her findings suggest that some species could swing their tails with enough speed that the perfect strike could generate bone-shattering force. They were highly effective weapons. A direct hit on a reckless rex leg might have equaled death for that individual. If these clublike weapons could be utilized for defensive purposes, does it imply that they could also be used as offensive weapons?

No, not to hunt down prey, bash it to death, and eat it. Ankylosaurs were strict herbivores. But perhaps for use during intraspecific combat. To help answer this question, first we need to look at a new species of ankylosaur described and named in 2017 by Arbour and David Evans, another ankylosaur expert from the Royal Ontario Museum in Toronto. This ankylosaur was discovered in 2014 near the city of Havre in northern Montana in rocks dating to a little over 75 million years ago. The partially mummified specimen is spectacular, exposing a giant shield of armor with skin overlying large spikes. Amazingly, it represents the most complete ankylosaurid ever found in North America.

Regarding the name, Arbour and Evans got creative and coined what must be one of the coolest dinosaur names out there, *Zuul crurivastator*. The genus was named after a memorable *Ghostbusters* movie monster, while the species name combines the Latin words for "shin" and "destroyer." This ankylosaur is the "Destroyer of Shins" because that is what its sledgehammer-like tail club would have done. The entire tail club, including the stiff handle, measures over 2 meters long. Destroying theropod shins might have helped it escape predation on occasion, but that is not all this club was good for. It might well have been a bit, well, sexy.

The evolution of the tail club in ankylosaurs like 6-meter-long *Zuul* has always been thought of as emerging as a defensive adaptation against predatory theropods. Conversely, this idea was challenged in a 2022 study by Arbour and Evans, joined this time by Lindsay Zanno, head of paleontology at the North Carolina Museum of Natural Sciences. Instead, they hypothesized that the tail clubs were, in fact, sexually selected structures that developed as a direct result of ankylosaur versus ankylosaur competition, or intraspecific combat.

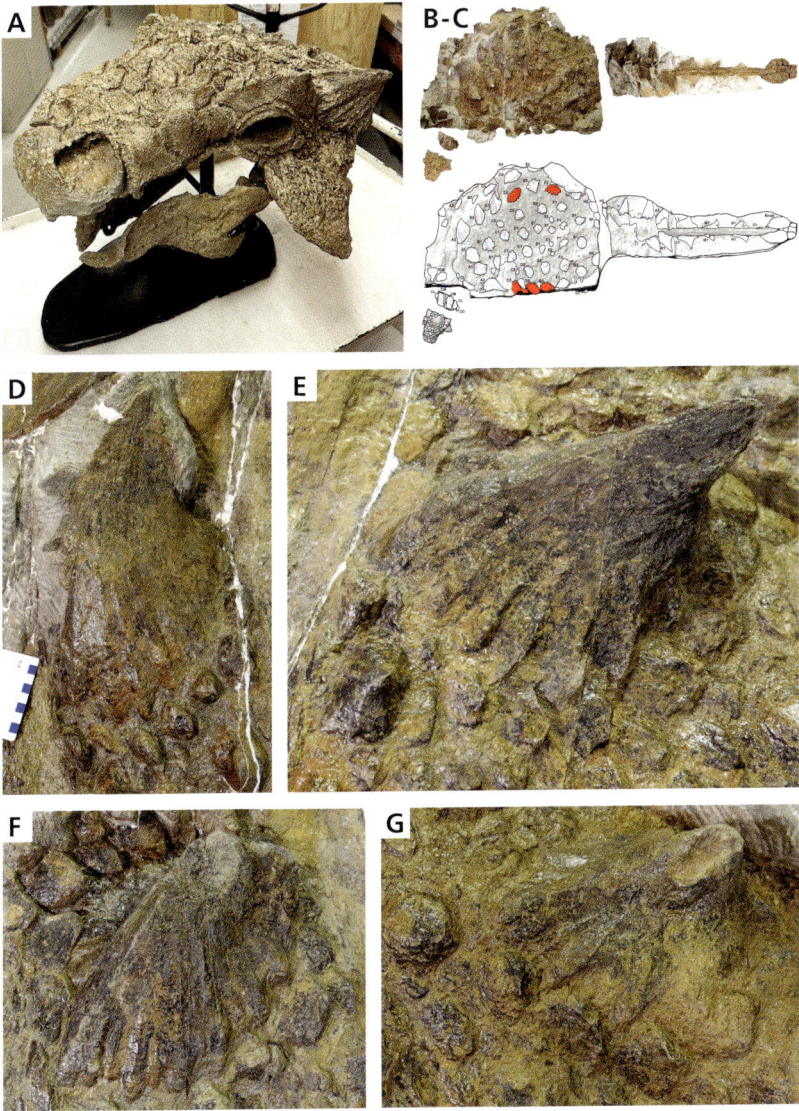

FIGURE 7.8. Various photos of the large ankylosaur *Zuul crurivastator*. (A) The exceptionally preserved skull. (B) Photograph and (C) interpretive illustration highlighting (red) the pathological osteoderms. (D–E) Two nonpathological osteoderms. (F–G) Two pathological osteoderms missing the tips; note that the keratinous sheath has not grown over the remaining tip.

([A] Photograph by the author; [B–G] courtesy of Danielle Dufault)

BOB NICHOLLS ART

When studying the remains of *Zuul*, the trio found evidence of gnarly pathological osteoderms concentrated solely in the hip region on both sides of the animal. Compared to the rest of the surrounding osteoderms, these pathological examples were dramatically different in shape and were missing their tips, although soft tissues (namely the keratinous sheath of the osteoderm) had grown over the broken area showing that it had healed. Moreover, the broken edges have a smooth surface texture indicative of reactive bone, suggesting some form of healing had occurred. The combined healing effects of smooth reactive bone and the keratinous covering are consistent with what is observed in living crocodilians with damaged osteoderms.

The pathological osteoderms appear to show different stages of healing. This might suggest that the individual engaged in multiple battles with contemporaries. It could also indicate that some of the wounds were infected and, therefore, took longer to heal. In any event, it is not a coincidence that the damaged osteoderms are positioned relatively symmetrically on either side of the animal. No similar or other definite pathologies could be identified elsewhere.

If these traumas were the result of an attack from a predator, say a tyrannosaur, then it could be expected that they would be more erratically placed across the body. Said tyrannosaur might take advantage of its far superior height and perhaps attack the more vulnerable neck, among other spots. Instead, these injuries are consistent with the impact zone of a swinging tail club delivering a smashing blow into an opponent's flanks. Considering that ankylosaurs like *Zuul* had a limited range of up and down movement in the tail, this suggests that a tail strike could only land at precise spots, including the head, shoulders, flanks, hips, and tail. The flanks, therefore, would have been in the range of the tail and were the perfect spot for another *Zuul* to target.

Similarly, some injuries in the pelvic area and on the tail club itself were briefly noted in a specimen of the Mongolian ankylosaur *Tarchia*

FIGURE 7.9. (Overleaf). *To the Victor Will Go the Spoils*

Two bull ankylosaurs, *Zuul crurivastator*, battle for dominance. Powerful blows from their tails target their bodies and heads—fragments of keratinous armor go flying!

tumanovae, possibly providing further support for injuries resulting from tail use during combat. As the team point out, direct comparisons with complete ankylosaurs are difficult because they are rare. Still, pathologies have yet to be reported in similar positions on those species of ankylosaurs that lack tail clubs.

Two rival *Zuul* smashing it out with their tail clubs should not appear extreme or out of the ordinary. It was probably one way to resolve hierarchies. Simply look to the modern animal kingdom and the diversity of herbivorous species that use weapons during intraspecific conflict; elephants clash with their tusks, deer smash with their antlers, and giraffes bash with their hornlike ossicones. The point is that display structures or weapons like these are often used in combat scenarios and may play an essential role in sexual selection.

Such fights are often ritualistic and reminiscent of tournaments rather than extreme death matches; otherwise, if an opponent died each time they fought, it would be a major disadvantage for the species. Still, such conflicts do often lead to injuries in specific areas directly associated with their weapons as observed in *Zuul*.

Although the evolution of these marvelous structures may have been driven by sexual selection, they served a dual role as visual deterrents and defensive weapons against would-be predators. Whether competing for dominance, territory, mating rights, or simply that last luscious leaf, the tail club played peacemaker among these tanklike warriors.

CLASH OF THE MIGHTY MARINE LIZARDS

Set and released twenty-two years after the original *Jurassic Park* movie and fourteen years after the last (*Jurassic Park III*), the eagerly anticipated sequel, *Jurassic World*, was unleashed in 2015. And with it came a flurry of new dinosaurs and other prehistoric animals for viewers to discover.

That is, if we ignore the fictional "Indominus rex," which admittedly quickly became a headache for many paleontologists because children and adults alike would state that it was their favorite dinosaur (I still occasionally hear this). Rather than arguing with a five-year-old who whips out their Indominus toy, you should politely explain that it was not a *real* dinosaur but something imaginary for the movie and that there are so many other cool choices. Still, I would prefer this to hearing them say that *Mosasaurus* was their favorite dinosaur! Yet another fan favorite, *Mosasaurus* was also not a dinosaur but a once very real giant marine lizard.

In *Jurassic World*, the *Mosasaurus* was depicted as some grossly oversized, 100+-foot-long gargantuan beast that was fed great white sharks for lunch. For sure, this exaggerated giant size is one of the reasons it became a firm favorite and why it stole the show in many ways. Alas, *Mosasaurus* was already cool enough, but hey, it is a monster movie and not a documentary. Top estimates put the real *Mosasaurus* around the 15- to potentially 17-meter mark. That is pretty enormous by marine reptile standards, making it one of the largest of all known species and dwarfing any modern reptile.

Mosasaurus lends its name to the wider group of famous aquatic lizards, the mosasaurs, which flourished during the last 30 million years of the Late Cretaceous. These mostly marine but occasional freshwater reptiles, whose closest surviving relatives are lizards and snakes, filled the apex predator niche in the oceans at a time when ichthyosaurs had gone extinct and plesiosaurs had passed their heyday. They were fully adapted to life in the watery world, with many species evolving body shapes convergent with whales and sharks.

Patrolling the ancient Cretaceous seaways, hunting down turtles, sharks, and other aquatic reptiles, mosasaurs surely came into direct conflict with each other occasionally, probably resulting in sometimes deadly consequences. Regardless of whether these interactions were between conspecifics (members of the same species) or other mosasaur species, they may well have fought over prey, territory, or dominance or even engaged in some form of aggressive courtship behavior when mating. We have some compelling evidence supporting such deadly duels of the deep. Strap in.

First, we go to the oceans of Kansas. Wait, there are no oceans in landlocked Kansas. What gives? During the Late Cretaceous, a huge, warm inland sea known as the Western Interior Seaway split North America in half and mosasaurs thrived here. One of the largest and among the most renowned was *Tylosaurus*, which reached about 12 to 15 m long. The first *Tylosaurus* fossils were described in 1869, and many more specimens have been found since, including skulls and skeletons from Kansas and other U.S. states.

In 2008, Michael J. Everhart, a renowned mosasaur researcher and adjunct curator of vertebrate paleontology at the Sternberg Museum of Natural History in Hays, Kansas, carefully examined the 70-centimeter-long skull, jaws, and cervical (neck) vertebrae of a beautifully preserved *Tylosaurus* in the museum's collections. This 85-million-year-old skull was found with multiple unmistakable puncture marks and gouges, including a puncture near the eye socket, two prominent gouges in the roof of the skull, and multiple marks on the right lower jaw.

At the time of Everhart's study, the skull was thought to have been a species that he had named a few years earlier, called *Tylosaurus kansasensis*. More recent work, however, has proposed that this species is likely the juvenile version of the first-named species, described in the 1869 study, *T. proriger*. In any case, based on the size of the skull compared with larger individuals, it was identified as a juvenile or subadult with a body length of 5 meters. On inspection of the injuries sustained to the skull, the deep gouges and puncture wounds were interpreted as unhealed bite marks. It seemed highly likely that this mosasaur died because of its injuries.

The only animal in the Western Interior Seaway with large enough jaws, a strong enough bite, and crushing teeth capable of delivering such a deadly blow was another mosasaur. Examination of the position and angle of key bite marks on the victim's skull roof showed that the initial attack probably came from the left and slightly behind, which almost certainly inflicted damage to the right eyeball, too. By measuring the distance between the

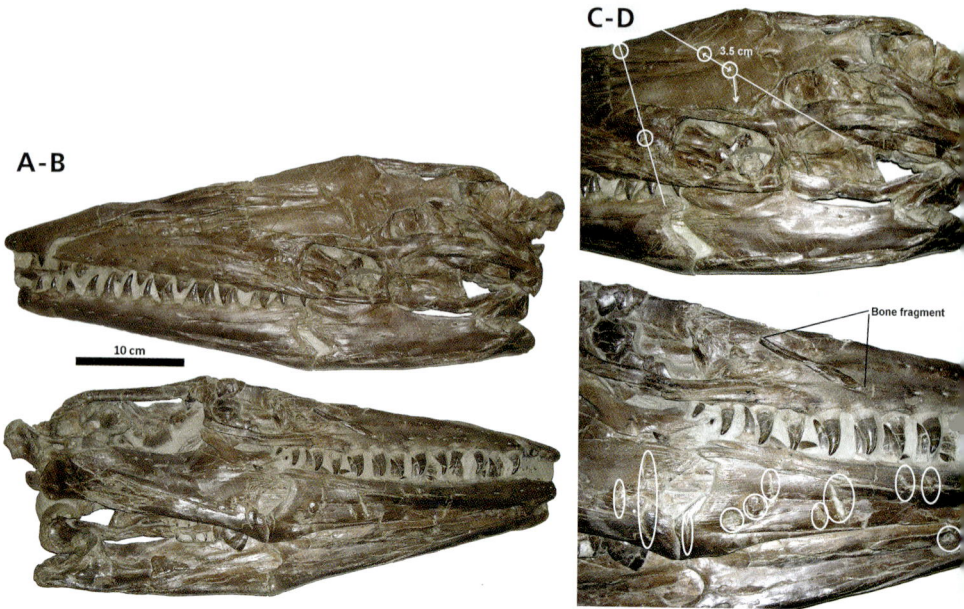

FIGURE 7.10. (A) The left and (B) right sides of the *Tylosaurus* skull described by Michael Everhart in 2008. (C–D) A close-up showing the various bite marks and gouges preserved across the skull. Lines indicate the approximate angle of the bite from the attacking *Tylosaurus*.

(Courtesy of Mike Everhart)

impact points of the teeth and comparing them with the distance between the teeth in the corresponding part of the jaw of a large *Tylosaurus*, Everhart estimated that the attacker might have been about 40 percent larger. That makes it around 7 meters long, representing a much more powerful, heftier animal.

Combined with the rest of the bite marks, the skull was fatally bitten and perhaps even partly crushed by the larger individual. The neck of the

FIGURE 7.11. (Opposite). *Eliminating the Competition*

A bite to the head of a smaller mosasaur, *Tylosaurus proriger*, gives the giant attacker the upper hand. When great sea serpents fight over territory, victory or death can be swift.

smaller mosasaur is also oddly angled, so much so that it might have been broken during the confrontation. Safe to say, the injuries suggest that this mosasaur probably died quickly.

In the same 2008 paper, Everhart provided an overview of what had been published up to that point regarding evidence for mosasaur battles. This included several specimens of different species found with, once again, skull and jaw fractures involving gouges and puncture wounds, many of which showed signs of healing. Some did not. Another displayed evidence of infection probably due to a bite, which led to the fusion of two vertebrae. The same individual also had bite marks on its tail, one on its left front flipper, and one on its jaws.

Since Mike's compilation, further examples have been reported. One such *Tylosaurus* from Texas, nicknamed "The Black Knight," had such a severe facial deformity that half of its face was missing. A chunk of the snout is absent, but evidence of healing in the form of bone remodeling is present and shows that it survived this dramatic situation. Part of the skull has possible tooth drag marks across its surface that seem to match those of another mosasaur, yet again rendering this a violent interaction between two mosasaurs. This survivor showed some serious resilience, capable of overcoming a life-threatening, lethal injury that probably eventually led to its demise. But, as hardy as this mosasaur was, it is nothing compared to what comes next on this deep-time journey . . .

8 | Different Diets

Chickens are typically fed a diet containing grains and seeds, with occasional treats of vegetables or fruits. But chickens are omnivores and sometimes show their instinctual theropod hunting behavior and attack, kill, and eat worms, bugs, and even mice that get too close.

MOSASAUR MASH UP

Ah, welcome back to the watery world of mosasaur wars. At the end of the previous chapter, we discussed some compelling evidence of a deadly mosasaur-eat-mosasaur attack, but there is more than meets (or should it be eats?) the eye with these mighty marine lizards.

When Everhart compiled the summary of research dedicated to mosasaur feeding behaviors, the most exceptional, standout specimens were those found with teeth still embedded in their victims. One of these comprised a skull and a partial skeleton of a *Tylosaurus* that was collected by him, which contained a large *Tylosaurus* tooth perfectly wedged between two cervical vertebrae. A specimen of *Mosasaurus conodon* from South Dakota was found with a tooth from a similar-sized individual pierced into the back of its skull. The bone around the embedded tooth showed no signs of healing, and the specimen also had other unhealed injuries including cuts and broken elements, along with a large puncture in the left lower jaw. As it turned out, the tooth's shape, structure, and size perfectly matched the teeth of *M. conodon*, confirming a *Mosasaurus*-on-*Mosasaurus* kill.

Painting a rather gnarly image of that last interaction, the skeleton of the victim *Mosasaurus* was found to be entirely disarticulated, with its bones scattered due to a scavenging feeding frenzy. Namely, more than three thousand shark teeth were found intimately associated with the skeleton, the bones of which also bear abundant fine cuts left by the sharks. Not only was this *Mosasaurus* killed by one of its kind, but its carcass was quite literally torn to pieces.

As somebody who has spent much of their career studying marine reptiles, I have examined a bunch of mosasaur fossils. One such fossil I got to see in 2017 was a skeleton on display at the Royal Tyrrell Museum in Alberta, Canada. This was yet another *Mosasaurus*, but this time probably of a species called *M. missouriensis*, which lived about seventy-four million years ago. The specimen was discovered by workers at the Korite Mine just outside of Lethbridge, in southern Alberta, and represents a fully articulated, 6.5-meter-long skeleton including the skull. Believe me, it is one of those beautiful fossils that sticks in your memory, but something else stuck with this mosasaur—a tooth.

The left lower jaw preserves at least three lesions; one of them is situated underneath the eye socket and contains a perfect tooth from, once again, a mosasaur of similar size. Like the previous example, the tooth is also characteristic of *Mosasaurus*, suggesting that we have another *Mosasaurus* versus *Mosasaurus* showdown. But this time the victim did not die. Each of the lesions shows signs of healing, with the presence of active bone remodeling, revealing that this individual survived the attack and lived to tell (and show) the tale with the embedded tooth in its face.

This discovery represents direct evidence of a nonlethal mosasaur attack on one of its own kind. Such intraspecific combat is known among living lizards, such as the Gila monster found in the southwestern United States and northwestern Mexico. When two males do battle, they bite each other, and the winner usually delivers the knockout blow by biting the opponent's throat region from underneath, stopping the rival from biting back. It is not unreasonable to presume that similar behavior might have been employed among this pair of fighting mosasaurs, especially when considering the position of the bite on just one side of the jaw and that the bite came from underneath. Such an attack must have been enough to subdue the other individual but not evidently enough to kill it. Rather than representing a predation attempt, this might suggest the pair clashed over territory or a mate.

This feeds nicely into direct evidence of mosasaurs eating mosasaurs. Not only do we have bite marks and embedded teeth, but we also have exceedingly rare fossils demonstrating that they consumed each other, too. A *Tylosaurus* found in Texas had the remains of at least three small mosasaurs in its gut, each belonging to a type called *Plesioplatecarpus* (then identified as *Platecarpus*). In one special last supper, a *Tylosaurus* from South Dakota was found to contain a mixed feast comprising parts of a *Latoplatecarpus* (also formally *Platecarpus*), a flightless seabird *Hesperornis*, and multiple fish. Contained inside yet another *Tylosaurus* from South Dakota were the partially digested remains of a different smaller mosasaur, called *Clidastes*, once again confirming that this top predator consumed smaller mosasaur species.

Most recently, in 2023, a remarkable mosasaur meal was revealed based on the discovery of a partial skeleton found at Bentiaba in Namibe Province, Angola. This mosasaur, identified as an adult *Prognathodon kianda* and estimated to have been about 7 meters long, lived toward the latest part of

FIGURE 8.1. (A) Skull of *Mosasaurus missouriensis* from near Lethbridge, southern Alberta. (B) A close-up of a tooth firmly embedded in the left lower jaw of the same specimen. (C) Prey items found inside the stomach region of *Prognathodon kianda* from Bentiaba, Namibe Province, Angola: green, *Gavialimimus*; yellow, cannibalized *P. kianda*; and pink, *Bentiabasaurus jacobsi*.

([A–B] Courtesy of Kathryn Abbott and the Royal Tyrrell Museum of Palaeontology; [C] courtesy of Michael Polcyn)

the Cretaceous, about 71.5 million years ago. Astonishingly, this individual contained three different, mashed-up mosasaurs inside its gut, each represented by a partial skeleton. One was an example of a long, narrow-snouted species called *Gavialimimus*, another an entirely new genus and species, *Bentiabasaurus jacobsi*, and the last a specimen of *P. kianda*. The latter is significant in that it confirms the first incidence of cannibalism in a mosasaur, a behavior previously lacking direct evidence.

FIGURE 8.2. (Opposite). *The Head-Hunter's Delight*

A hungry mosasaur, *Prognathodon kianda*, enthusiastically swallows the decaying head of another *Prognathodon kianda*. This is his hat trick, a third mosasaur meal in a row!

Each mosasaur meal had been dismembered to some extent and measurements of the remains show that they fall between 43 and 57 percent of the hunter's body length. Curiously, there are differences in preservation among each prey item, with those remains in the foregut region (encompassing *Bentiabasaurus*) showing little or no erosion, but the bones in the hindgut (*Gavialimimus*) display strong evidence of corrosion from stomach acids; for instance, most of the tooth crowns had dissolved away.

As might be expected, this indicates that the predatory *Prognathodon* consumed each meal on three separate occasions. It is rather peculiar, however, that each of the mosasaur meals is represented mainly by skull elements. Maybe this predator was some strange, specialized headhunter? Targeting the head and consuming just that would be highly suspect given that the skull is the boniest part of the body and one of the least nutritious, too, not necessarily the ideal food source for a massive mosasaur. It might be coincidental or maybe some odd preservation bias, but it raises the possibility as to whether this mosasaur scavenged these meals instead, which would appear more likely. Maybe it ate the leftovers that a larger *Prognanthodon* had previously killed and consumed. Either way, like many predators, it was probably an opportunist and would have opted for a free meal to save energy and reduce the risk of injury.

Head biting, tooth embedding, and mosasaur munching—the evidence presented among these incredible fossils shows that mosasaur versus mosasaur contests must have been quite common encounters. In some cases, these conflicts would leave one opponent with battle scars. Others were not so lucky. Losing a fight could mean losing your life, perhaps even becoming the next meal of your giant marine lizard adversary.

MICRO MEALS—FOOD FOR THOUGHT

One of the most famous and iconic of all dinosaurs to come out of China over the last thirty years must be *Microraptor*. Though things did not start so well for this dino icon, as the first specimen was found to be part of a fossil forgery made up of three separate dinosaurs (sigh), its scientific significance was eventually realized, and over three hundred fossils have since been found, all from the roughly 120-million-year-old Early Cretaceous rocks of the Jehol Biota in China's Liaoning Province.

With its amazing four wings, two on the arms and two on the legs, *Microraptor* has become a well-studied birdlike dinosaur and a staple focal point in research on the evolution of early avian flight. To give it some context, *Microraptor* belongs to the dromaeosaur (or "raptor") family of dinosaurs, the same wider family that includes famous types like *Velociraptor* and *Deinonychus*, but it belongs to its own group within that family called Microraptoria. As of today, three species are identified, including *Microraptor gui*, described in 2003, which is my personal favorite because the holotype is so visually spectacular and complete with all wings in place. Plus, in 2017, I finally got to see the real specimen on display as part of a special "Dinosaurs of China" exhibition at Wollaton Hall Natural History Museum in Nottingham, UK.

At about the size of a large crow or raven, with a maximum wingspan of around a meter and sporting a long tail, *Microraptor* is one of the smallest known nonavian dinosaurs. Studies have shown that this little theropod was a comfortable glider and was probably also capable of powered flight. But what did it eat? Was this dromaeosaur a mini version of its famous carnivorous cousins, or did it eat seeds and fruits like the contemporaneous fruit gulping *Jeholornis*? With lots of *Microraptor* fossils, many preserving feathers and some with soft tissues, it would seem there is a fair chance that at least one of them would reveal evidence about this dinosaur's diet.

As it turned out, the first direct evidence of diet was particularly intriguing because one specimen of *M. gui* was found with the adult remains of a bird inside its stomach cavity. This was unique, direct evidence of a dinosaur-on-dinosaur predator-prey relationship. The type of bird was a toothy enantiornithine, a group of birds that went extinct at the end of the Cretaceous, and this one was consumed by this *Microraptor*. The study was published by a trio of paleontologists in 2011, led by Jingmai O'Connor, known as the "punk rock paleontologist," now a curator at the Field Museum in Chicago, who has studied and described her fair share of Chinese birdlike dinosaurs. The famed Chinese paleontologists Zhonghe Zhou and Xu Xing made up the rest of the team, the latter being one of the most prolific paleontologists on the planet, who has named more dinosaurs than any other living paleontologist. Unsurprisingly, he originally named and described *Microraptor*, too.

Contained inside the gut of this theropod is the undigested partial skeleton of its bird dinner, including parts of a wing and both feet with associated claws. The remains are still in articulation, suggesting perhaps that this was not the result of scavenging a few isolated bits, and it appears the bird was swallowed headfirst. For size comparison, *Microraptor* is estimated to have weighed about 1.5 kilograms, whereas this little bird is thought to have been 60–70 grams. Digestion had probably not progressed significantly, and this suggests that the *Microraptor* died a little while after consuming the bird.

This discovery also plays a role in understanding the lifestyle of this mini predator within its forested, freshwater lake environment. Studies of enantiornithines from this formation show that they were arboreal, and this discovery provides some support for the arboreality of *Microraptor*. It may show that this dromaeosaurid spent at least some of its time hunting in trees, perhaps gliding from tree to tree in hot pursuit of its prey.

A couple of years later, in 2013, another study focused on yet another specimen of *M. gui* with its last meal preserved. This time, the Mesozoic meal was something that might have been a little more expected: fish, which are the most common vertebrate fossils from the Jehol Biota. The dentition of *Microraptor*, with forward-projecting teeth and reduced tooth serrations more useful for spearing, is consistent with a predator that, at least partially, would have been piscivorous. The gut contents are marked by a dense oval mass containing numerous disarticulated bone fragments comprising fin rays, ribs, fragments of spine, and cranial bones from numerous fish. Some of them display acid etching. Unless these fish lived in trees, the discovery of fish inside the stomach shows that *Microraptor* did not strictly hunt in an arboreal environment and consumed aquatic prey, too.

In 2019, O'Connor led a team that described a new specimen of the type species *Microraptor zhaoianus*, the first species originally described in 2000, which contained a practically complete lizard inside its stomach. Furthermore, this lizard lunch turned out to be an entirely new species that the team called *Indrasaurus wangi*. Imagine that? The lizard's death was not in vain. Instead, the capture and subsequent consumption of the lizard is what actually led to the fossilization and eventual recognition that this species was ever even present on the planet and was thus recognized as a new species inside the stomach of the animal that ate it.

As the team discussed, this *Indrasaurus* was largely complete and well-articulated, including its head, body, tail and feet, reconfirming that *Microraptor* was an agile predator that, in this case, probably snagged its prey and ingested it whole and once again headfirst. This feeding position is consistent with what is found in living carnivorous birds and lizards, which usually consume their prey in this headfirst position.

Microraptor meal number four came in late December 2022. This Christmas cracker of a rediscovery was revealed following reassessment of the holotype of *M. zhaoianus*, which turned out to contain a perfectly preserved right foot of an early micromammal inside its stomach! Now, technically, some of the academic community already knew about this mammal meal because of a presentation and a published abstract for a Society of Vertebrate Paleontology conference in Pittsburgh in 2010, but the formal description had yet to appear in print. It is rather strange that the pretty obvious foot had previously been missed by so many researchers, but in many ways, this is understandable considering that your mind would not necessarily be trained to focus on looking for mammals—especially their feet—inside dinosaurs, right?

In any case, this is a rare definite example of a dinosaur eating a mammal. A pretty awesome find all round. As the mammal is only represented by an isolated foot—at least, that is all that can be seen—the team were unable to provide a positive identification, although one thing for sure is that it is from an animal about the size of a mouse. Moreover, the foot does not seem to suggest any sort of unique adaptation for climbing, so it perhaps lived on the forest floor where it was presumably picked up (and scavenged) by the *Microraptor*.

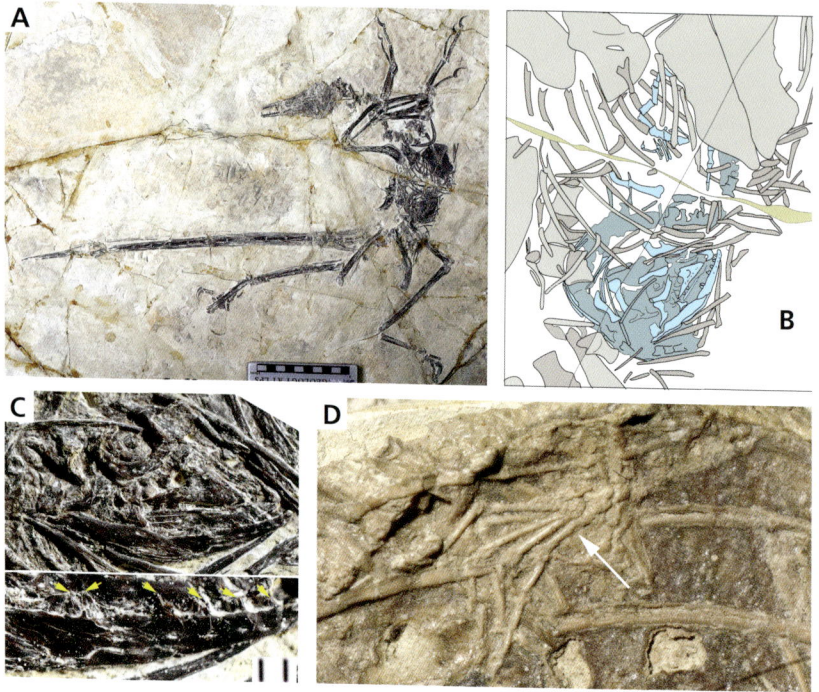

FIGURE 8.3. (A) Photograph and (B) interpretive illustration of a *Microraptor zhaoianus* with a lizard (*Indrasaurus wangi*) dinner. (C) A close-up of the lizard's skull and part of its jaws; arrows point to some of the teeth of the lizard. (D) Arrow points to an isolated mammal foot found inside the stomach of a *M. zhaoianus*.

([A–C] Courtesy of Jingmai O'Connor; [D] courtesy of Alex Dececchi and Dave Hone)

Choices, choices. *Microraptor* could really put it away and seemed to have eaten whatever it could catch or scavenge. This is typical of what we think of, that many predators are opportunistic and generalist feeders. It is near impossible to say 100 percent whether any of the four micro meals represent definite predation or scavenging, but they do at least confirm that *Microraptor* had a highly varied diet of birds, fish, lizards, and mammals.

FIGURE 8.4. (Opposite). *Arboreal Fine Dining*

Microraptor gui, a feathered dinosaur with wings on both its arms and legs, greedily gulps down a lonely lizard, *Indrasaurus wangi*.

BOB NICHOLLS ART

Quite the menu for this dromaeosaurid, seemingly an agile little hunter in trees, on the ground, and along rivers and lakes. Nobody was safe.

As much of a tiny terror *Microraptor* might have been, with small size comes bigger predators. In the same environment, other larger theropods are known and have been found with smaller dinosaurs in their stomach, like the much larger, multimeter-long *Sinocalliopteryx* with one fossil found to contain a dromaeosaurid leg (belonging to a *Sinornithosaurus*) in its stomach and another with the remains of two early birds (*Confuciusornis*) inside. Plus, we also have dinosaur-eating mammals that could easily have eaten a *Microraptor*. That said, to my knowledge, there is no direct evidence of any *Microraptor* as the final meal of any contemporary predator. Perhaps time will tell. Still, for now, with such a diversity of delicacies, we know more about the diet of *Microraptor* than any other dinosaur.

FISH ARE FOOD, NOT FRIENDS

"Dr. Lomax, what is your favorite dinosaur?" is a question I can almost guarantee will come up in any Q&A I am involved in. People love to ask this question, and it usually comes from children who are so excited to hear my answer and why. For an adult, being asked what your favorite dinosaur is can feel somewhat nostalgic. This is why the science communicator Jimmy Waldron, from the Dinosaurs Will Always Be Awesome traveling museum, who is also my good friend, literally asks *everybody* this question, and the response is always met with tremendous excitement, clearly unlocking some early core memories. As a child, my favorite was *Stegosaurus* simply because of the unusual plates and spikes, but as I entered my teen years, my favorite shifted and has forever since remained the same: *Baryonyx*.

I apologize to *Stegosaurus* and other herbivores, but the size and big predator status of *Baryonyx* is what initially attracted me to it, although its discovery and significance are what kept me in camp *Baryonyx*. Let me explain. *Baryonyx* was originally found in a clay pit called Smokejacks in Surrey, England, in January 1983 by the plumber turned fossil collector William Walker. He initially found part of what turned out to be a ~30-centimeter claw and informed the Natural History Museum, which sent a team to excavate what turned out to be a pretty complete skeleton of what was then the largest theropod dinosaur known in Europe, reaching around 9–10 meters in length. The discovery was even dubbed the "find of the century" because of its importance, especially in that no other reasonably complete large theropod had previously been found in Early Cretaceous rocks anywhere and because it was just so different. It was formally announced as a remarkable new dinosaur in the prestigious journal *Nature* in 1986, many years before my career began. In fact, before I was even born.

One of the most intriguing things about this dinosaur, and what grabbed my attention, was that it had its last meal preserved. Contained inside the stomach of this 130-million-year-old, Early Cretaceous predator, were a couple of bones from a juvenile iguanodontid (*Iguanodon*-like) dinosaur. This was fascinating because it revealed that *Baryonyx* ate baby dinosaurs, though the most significant finding was that of fish remains also inside the gut. If you are a long-time dino fan, you have probably heard of the discovery of *Baryonyx* and know that it ate fish, but the story and its significance are often underplayed, which is surprising considering that this was the world's first confirmed fish-eating dinosaur and changed the dino diet world forever.

With its long and very narrow snout packed with croc-like conical teeth, it was assumed that *Baryonyx* spent time in the water and might have been piscivorous. When the skeleton was being carefully cleaned in the Natural History Museum's laboratory, a portion containing the ribs of *Baryonyx* revealed a mass of partially digested fish scales and teeth, clearly representing part of the dinosaur's last meal. They were initially identified as belonging to a small to moderately sized fish called *Lepidotes*, although more recent classification has shown that it belongs to a type called *Scheenstia*. A fishing *Baryonyx* is thought to have used its large claws to gaff fish out of the water and to snatch up slippery prey with its long, gharial-like jaws.

As part of my research as a paleontologist, and specifically for a book that I wrote earlier in my career, *Dinosaurs of the British Isles*, I had the pleasure of studying the bones of *Baryonyx* and, yes, even taking a moment or two to handle the infamous claw. In doing so, I also had a chance to inspect the fish remains from inside the gut, which shows evidence of acid etching from stomach juices during the digestion process. The juvenile iguanodontid also shows some evidence of abrasion, probably from stomach acids, further confirming the last meal.

The discovery of fish remains helped put two and two together and established the fish-eater identity, which raised new questions for our understanding of spinosaurids, the types of dinosaurs to which *Baryonyx* belongs. You will no doubt have heard of the bizarre, sail-backed *Spinosaurus*, which is the poster child for this group of dinosaurs. There has been a lot of work, especially recently, dedicated to understanding spinosaurid ecology and whether they spent much or even most of their time in the water, dining primarily on fish. For my sanity, and yours, I do not wish to get into the ins and outs of *Spinosaurus* here because there has been a lot of conflicting research focused on that dinosaur lately, with more to come, but there is no doubt that it spent a chunk of time in the water where it fed on fish, such as car-sized coelacanths that have been found in the same rocks.

For spinosaurids generally, they are presumed to have spent some time in the water, be it wading for food or loitering around the edges of rivers and lakes, and to have had a diet of mostly fish. Emphasizing the significance of *Baryonyx*, to this day, aside from some possible isolated teeth, the skeleton remains the only known definite example of this dinosaur in the world, and it is the only spinosaurid to have direct evidence of its last meal of fish and a dinosaur inside its stomach. Though it is not the only spinosaurid to provide direct evidence of diet.

In 2004, a short study was published that sparked a lot of interest in the public domain. The focus was on a group of three cervical (neck) vertebrae belonging to a pterosaur from the Early Cretaceous rocks of the Santana Formation in northeastern Brazil. Based on the size of the vertebrae, it was estimated that the pterosaur had a wingspan of about 3.3 meters. As nice as the vertebrae are, the focus was on a broken spinosaurid tooth embedded on the right side of the anterior-most vertebra. This tooth is similar in every way to known spinosaurid teeth from the Santana Formation, which has yielded the spinosaurid called *Irritator*. In this situation, a spinosaur,

FIGURE 8.5. (A) The author holding the original thumb claw of *Baryonyx walkeri*. (B) Fish scales from the stomach contents.

(Photographs by the author)

perhaps an *Irritator*, had chomped on what must have been a relatively fresh and articulated pterosaur neck, only to break a tooth in the process. It is hard to say whether the spino caught the pterosaur while it was scrambling on the ground or even in the air, though it would seem more likely that this is an example of scavenging. Evidently, given the articulation and lack of evidence of stomach acids, the vertebrae were not swallowed.

By looking at the shape and structure of their skulls and jaws, combined with multiple studies and the presence of fish in *Baryonyx*, it is evident that *Baryonyx* and kin fed largely on a diet of fish. The fact, however, that a juvenile iguanodontid was found inside the stomach of *Baryonyx* and a spinosaurid tooth was embedded in a pterosaur indicates that spinosaurids were opportunistic, too, feeding on whatever might come their way. The discovery of *Baryonyx* was sensational, not only because of how unusual this dinosaur was but also how it helped change our understanding of theropod dinosaur diets and provide new ways of thinking about dinosaur ecology. This, my friend, is why *Baryonyx* will forever be my favorite dinosaur.

FIGURE 8.6. (Overleaf). *Rulers of the Beach and Shallows*

A juvenile ornithopod dinosaur, *Mantellisaurus atherfieldensis*, runs for its life from an opportunistic *Baryonyx walkeri*, a spinosaurid dinosaur. Two other *Baryonyx* are successfully fishing in the river.

SEA DRAGON SUPPER SURPRISE

Diving into the niche world of ichthyosaur stomach contents, the early literature from the 1800s reveals that ichthyosaurs with their last meals preserved have been recorded from several locations. The discoveries show these reptiles had a taste for fish and squid, and there were often remnants of fish scales, bones, or cephalopod arm hooklets preserved in their guts.

As you may remember from the introduction, my first-ever study focused on an ichthyosaur that preserved stomach contents comprising squid and fish, the one we later named *Ichthyosaurus anningae*. Well, prior to this study, the preceding description of an ichthyosaur from the UK with its last meal intact was reported in 1968. That one contained three or four different types of squid hooks, possibly from different species. The stomach contents of British ichthyosaurs were apparently rare, but the 1968 study showed that quite a few specimens with their last meals had been overlooked.

In this vein, a study in 2021 examined the stomach contents of the Early Jurassic ichthyosaur *Stenopterygius quadriscissus* from Holzmaden, Germany. By examining the stomach contents of juvenile and adult ichthyosaurs, changes in tooth shape were found to be associated with age and prey preference. The team found that young *Stenopterygius* mostly fed on small, burst-swimming fish, and slightly older individuals transitioned to feeding on faster-moving fish and squid-like cephalopods. In contrast, fully mature adults appear to have dined exclusively on cephalopods.

Undoubtedly, squid and fish were abundant and high on the ichthyosaur menu, presumably throughout their span in the Mesozoic, making up a significant part of their diet. This can be expected for ichthyosaurs, which we can also compare with modern-day dolphin and whale analogues, which primarily eat the same. However, just like dolphins and whales, whose diets are not entirely uniform, several ichthyosaurs have been found to have bizarre last meals.

Cannibalism is the first thing to mention. Some ichthyosaurs, especially from the Holzmaden area, have been found with smaller ichthyosaurs inside that represent food rather than embryos (ichthyosaurs gave birth to live young and many gravid females are known). One of the perhaps most gruesome examples is of an aborted, early-stage embryo that was found within a gastric mass of cephalopod hooklets inside an adult *Stenopterygius* that was itself pregnant with a late-term embryo. In these Early Jurassic seas, large-bodied ichthyosaurs like *Temnodontosaurus* are considered the apex predators and no doubt fed upon various animals. One example of an 8-meter-long individual, also from Holzmaden, has cephalopod remains and as many as four juvenile *Stenopterygius* inside its stomach that were preyed upon at different times, perhaps days apart.

The earliest example we have of an unusual ichthyosaur diet comes from Middle Triassic rocks of southwestern China. At about 240 million years old, the spotlight fossil comprises a nearly complete skeleton of an ichthyosaur called *Guizhouichthyosaurus* that was unearthed in 2010 at Xingyi, Guizhou (hence the name). This type is known from multiple individuals and is not what you might generally think of as an apex predator, especially compared with later ichthyosaurs with massive skulls, large cutting teeth, and more robust bodies, such as *Temnodontosaurus*. As the old saying goes, looks can be deceiving.

Inside the abdominal region of this 5-meter ichthyosaur is a large, bulging mass packed full of the undigested bones of another marine reptile called a thalattosaur, specifically a species named *Xinpusaurus xingyiensis*. This is intriguing because it confirms that these early ichthyosaurs dined on other marine reptiles, but more significantly, this thalattosaur would have measured a whopping 4 meters long!

The teeth of Guizho are perfect for grasping prey, and it appears, based on the research presented in the 2020 study, that it swallowed the torso of the prey in what might have been just one large piece, or perhaps a few sections. The skull and tail were not present, although an isolated tail probably from this individual was found 23 meters away from the skeleton. If the tail belongs to the prey, then logic suggests that the predator must have died soon after ingesting its oversized meal; the lack of acid etching on the bones in the stomach is also in agreement with this interpretation. For several reasons, this very likely represents predation over scavenging, but notably, if another predator had killed the *Xinpusaurus*, it would be unusual for the predator to have left behind the most nutritious bits of the trunk and limbs. With two large-bodied individuals, this act of megapredation likely represents the oldest direct record of megafaunal predation among marine tetrapods, in this case of a large predator feasting on similarly sized prey.

Saving the strangest till last, we move forward in time to the Early Cretaceous and to the ~105-million-year-old partial skeleton of an ichthyosaur collected in Queensland, Australia. In 2003, a short study focused on this ichthyosaur, a type known as *Platypterygius*, and provided the most extreme evidence of an ichthyosaur diet to date. Contained inside the gut of this roughly 7-meter ichthyosaur are fish remains and, somewhat incredibly, hatchling-sized turtles and even a bird! This was and still is the first and only known evidence of feeding by ichthyosaurs on both turtles and birds.

The hatchling turtles belong to an extinct family known as protostegids, the same family that includes the largest turtle ever to have lived, *Archelon*; they lacked the classic, hardened shell and had more of a leathery shell, like modern-day leatherback turtles. Unfortunately for the tiny 10–12-centimeter snacklings, their presumably poor swimming abilities may have made them easy targets, with the ichthyosaur perhaps swallowing them whole or shaking them from side to side and ripping off smaller pieces during ingestion.

The bird is represented by part of a leg (a portion of a tibiotarsus) from an adult and can be readily identified as that of a toothed type known as an enantiornithine, perhaps a type called *Nanantius eos*. This particular individual went extinct in the stomach of the ichthyosaur, but whether it was snatched by the ichthyosaur or was already dead, floating on the water's surface, is impossible to say. On one hand, the more romanticized idea of an ichthyosaur darting out of the water and snatching a bird flying above seems unlikely, but not impossible, though perhaps a more plausible interpretation for active predation is that the bird might have been sitting on the surface and was taken by surprise, much like modern-day seabirds taken by fish and cetaceans. Most likely the bird was already dead and floating with the waves, but you never know.

This is my personal favorite of all known ichthyosaur stomach contents. Yeah, I know that it is weird to have a favorite type of long-dead creature's final meal, but that is just how nerdy you can get in paleontology. I remember reading about this when undertaking the research for my first study.

A

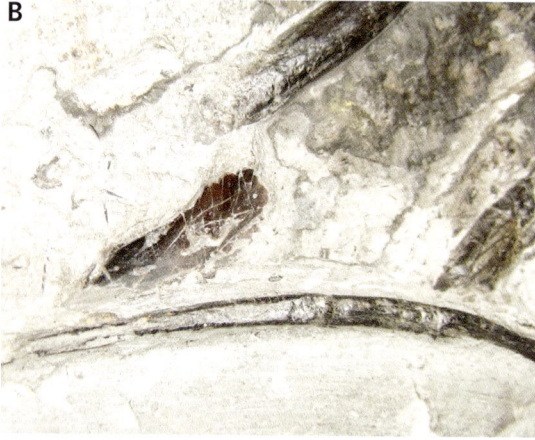

B

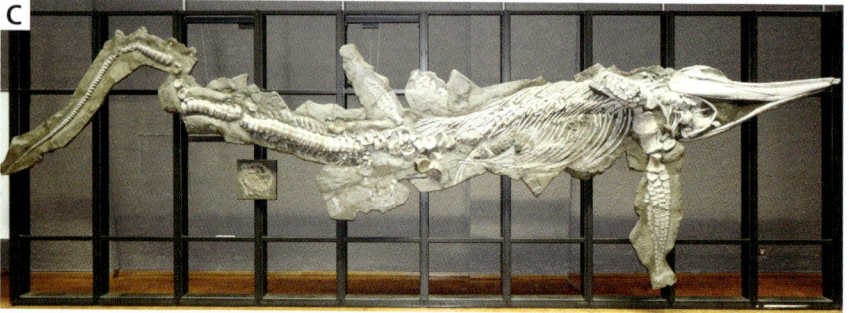

C

D

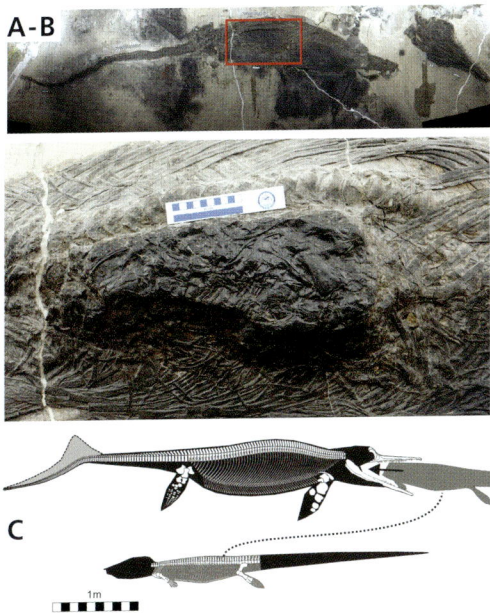

FIGURE 8.8. (A) The complete skeleton of *Guizhouichthyosaurus* from the Middle Triassic of Xingyi, Guizhou, China. (B) Stomach contents containing the thalattosaur, *Xinpusaurus xingyiensis*. (C) Interpretive illustration showing the ichthyosaur consuming the thalattosaur.

(Images from D-Y. Jiang, et al., "Evidence Supporting Predation of 4-m Marine Reptile by Triassic Megapredator," *iScience* 23 (2020): 101347)

FIGURE 8.7. (Opposite). (A) The holotype of *Ichthyosaurus anningae* from the Early Jurassic near Lyme Regis, Dorset, England. Note the black mass (stomach contents) between the ribs. (B) A fish scale and example hooklets from the arms of squid contained inside the stomach contents. (C) A *Temnodontosaurus trigonodon* skeleton from the Early Jurassic of Holzmaden, Germany. (D) A close-up of part of the stomach contents of (C) containing juvenile *Stenopterygius* vertebrae.

(Photographs by the author)

BOB NICHOLLS ART

What makes it extra special is the fact that the ichthyosaur was also pregnant and close to parturition. It made me ponder whether this ichthyosaur, much like pregnant women who may get a sudden craving to eat a specific type of food, also had the urge to consume other animals that would not typically be part of its diet.

Being pregnant, the expectant mother would require more energy to fuel herself and her unborn young, so perhaps there is something in this. Or perhaps this species dined on more varied prey and just so happened to preserve this unique snapshot. The researchers noted that this ichthyosaur is closer in geological time to when the group finally bowed out, a little before ninety million years ago, and so perhaps the unusual diet might reflect the result of having more competition for their typical cephalopod prey, thus forcing them to hunt elsewhere. Sadly, for the ichthyosaur, the apparent lack of good evidence for the digestion of the multipart meal suggests that the gravid female died shortly after she consumed her last supper.

Outside of the examples discussed herein, several ichthyosaurs with mostly cephalopod or fish suppers have been documented at various sites around the world. Having numerous examples of ichthyosaurs with direct evidence that they feasted on squid and fish show they were a favored food item. The much rarer and, dare I say, cooler findings that ichthyosaurs ate other ichthyosaurs, as well as thalattosaurs, baby turtles, and even birds, demonstrate that these marine reptiles were not fussy eaters and were more opportunistic than previously thought.

FIGURE 8.9. (Opposite). *The Hungry Hungry Ichthyosaur*

Out of the blue, a predatory ichthyosaur (*Platypterygius*) approaches a bale of baby turtles. They will be easy pickings, and so will the bird (*Nanantius eos*), which is oblivious to the danger.

9 | Digestion

To rid their stomachs of unwanted food and do a little spring cleaning, some sharks can rapidly puke their guts out before retracting them again. This bizarre feat of gastrointestinal engineering is known as gastric eversion.

A COPROLITE FIT FOR A KING . . . OR MAYBE A QUEEN

Casually walking over to a mountain of dino droppings, charismatically pulling off his glasses to take a closer look and exclaiming, "That is one big pile of shit," Dr. Ian Malcolm, played by Jeff Goldblum, gave us a memorable line and rather vivid image in *Jurassic Park*. If you recall, the sick *Triceratops* had apparently made two enormous poop piles, one taller than Goldblum, but the reality, however, is that no single coprolite this size has ever been discovered. I always liked this scene, though, because it got people thinking about dinosaur poop and the very fact that to understand what the animal had been eating, and why it might have been sick, you needed to study the droppings, as Dr. Ellie Sattler did so tenaciously.

Laura Dern's character was certainly correct to get so hands on because, to be straight, poo says a lot about your health and diet and is a good place to start if you want to learn about an animal's feeding behavior. This latter point is a nice link to paleontologists studying fossil feces (coprolites). It is also nice that it was a female character studying the *Triceratops* droppings because the first person to bring fossil feces to science was none other than the pioneering paleontologist Mary Anning.

For a long time, coprolites were known as "bezoar stones," and Anning noticed that they were sometimes found inside the abdominal region of ichthyosaurs and that isolated specimens contained fish remains. She passed this information onto the leading geologist William Buckland, perhaps most famous for naming the first dinosaur, *Megalosaurus* (in 1824), who also coined the name coprolite in 1829 for these poop stones. Buckland often credited Anning and mentioned her by name in this study, giving her full credit for both her skill at finding these fossils and for her identifications. On a neat little side note, I currently have one of their 1829 coprolites in my office, which I am studying as part of an ongoing research project. This one contains the bones of a baby ichthyosaur.

In the world of fossil poo today, one name always quickly comes up in conversation, Karen Chin. It's no slur to say her research is poo; rather, she is a leading world expert on fossil feces. Chin is a professor and curator at the University of Colorado Boulder and has published extensively on the subject. As a paleontologist, or what one might call paleocoproliteologist (is that a thing?), she and her team(s) have done some genuinely pioneering research on fossil poop.

On a personal connection, in my first research paper, the one that looked at stomach contents in the ichthyosaur, I remember identifying a mass of what looked like a coprolite associated with the ichthyosaur skeleton. I reached out to Chin, who quickly got back to me to confirm my suspicions. Over the years, I have spent time studying other coprolites and have been known to hand out a mystery fossil during my lectures, only to reveal later that it was a fossilized poop. It is always amusing to see the children (and adults!) sniff and wipe their hands afterward.

Beyond stomach contents, coprolites may provide some of the best information about an animal's eating habits, at least if we are lucky. Chin has undertaken a lot of research focusing on dinosaur diets through evidence from coprolites. One of her most notable contributions, and unarguably one of the most famous coprolites, was reported in 1998 when Chin and her colleagues published a study in the journal *Nature* that focused on a very large dinosaur coprolite.

Channeling the same energy as the giant *Triceratops* movie poop, by human nature, we tend to have a fixation with the "biggest" or "heaviest" and like to add titles to things, so finding the biggest dinosaur poo would be quite a feat and would definitely turn a few heads. The coprolite in question was discovered in Chambery Coulee valley in southwestern Saskatchewan, Canada, and measured a record-breaking 44 centimeters long, 13 centimeters high, and 16 centimeters wide. The original, freshly deposited dung would have been slightly larger, considering that its mass would have been altered through drying, compression, and so forth during the fossilization process.

Compared with a Jeff Goldblum–sized poo, this might not seem all that amazing, but it is genuinely huge, more than twice as large as any previously reported carnivore coprolite at the time. Moreover, this was not reported to be any old dinosaur coprolite, but as the title of their 1998 study revealed, this was "a king-sized theropod coprolite." The "king" alludes to *Tyrannosaurus rex*, the "tyrant lizard king." Yes, this was almost certainly a *T. rex* coprolite. But hold on to your *T. rex* butts, how could they know for sure that a *T. rex* made this coprolite?

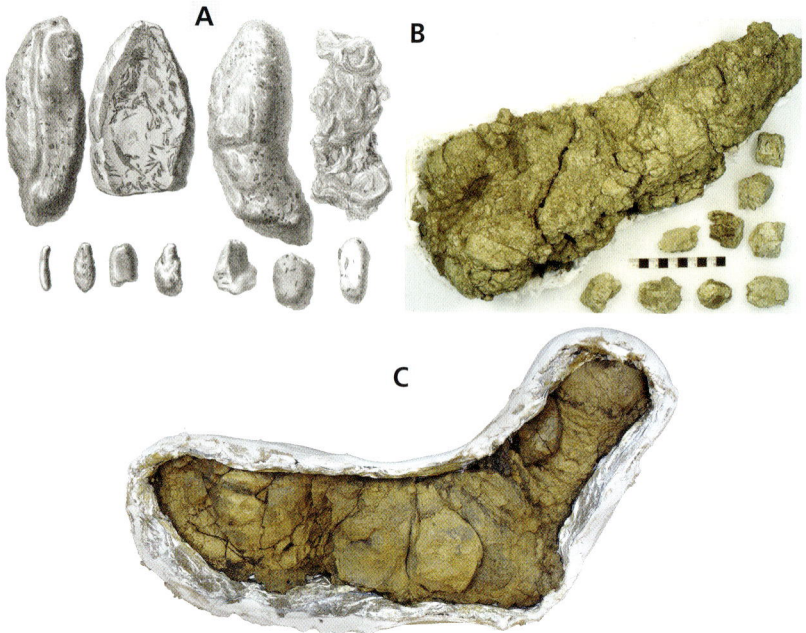

FIGURE 9.1. (A) Numerous coprolites, including many found by Mary Anning, described and illustrated by William Buckland in 1829. Several contain the remains of squid hooklets. (B) The large *Tyrannosaurus rex* coprolite from southwestern Saskatchewan, Canada. (C) "Barnum," the second *T. rex* coprolite, was found in South Dakota.

([A] Image from Buckland, W. 1829. See references; [B] courtesy of Karen Chin, this coprolite is a part of the paleontology collection of the Royal Saskatchewan Museum; [C] courtesy of George Frandsen and the Poozeum)

It should go without saying that *Tyrannosaurus* would have pooped a lot in its lifetime. Depending on the perfect conditions for fossilization, there would always be a chance that at least one of its droppings would work its way into the fossil record. Still, when Chin's team published their research, they explained that despite the thousands of known coprolites, specimens that could be unequivocally identified as having been made by a carnivorous dinosaur were virtually unknown.

Assigning a coprolite to its maker is an inherently difficult task, but it can be done when the coprolite is found in the same-aged rocks and same geographic area as the potential producer. Our dinosaur coprolite in question is from a 66-million-year-old Late Cretaceous (Maastrichtian) rock formation known as the Frenchman Formation. Here, *T. rex* specimens have been found, including "Scotty," a candidate for the largest *T. rex* specimen in the world. It is coeval in time with the famed Hell Creek Formation. That is the first connection.

Diving deeper into this fossil scat, it contains (up to 30–50 percent) large fragments of munched-up, dark brown broken bones from a subadult dinosaur. Clearly, this coprolite was made by a carnivorous dinosaur and strongly suggests that a *T. rex* produced it, the only large-bodied theropod from this formation that would have been capable of producing such a sizable prehistoric poo. For reference, much, much smaller theropods (and crocs) have been found in this formation, such as small ornithomimids and dromaeosaurs, but nothing close to the size of an animal capable of creating such a coprolite. This is the same level of detail and process of elimination that the research team discussed when examining the large coprolite. Though we can be almost certain that a *T. rex* made this Cretaceous coprolite, we will never know whether it was a male or female.

A second specimen of what is almost certainly another *T. rex* coprolite was found in 2019. This time, the specimen came from the Late Cretaceous Hell Creek Formation in South Dakota, famed for *T. rex* fossils, and was acquired by the Poozeum. Yes, really. This museum was founded by the coprolite fanatic George Frandsen, who has amassed a colossal coprolite collection, parts of which are often loaned to museums across the United States. The museum is based in Williams, Arizona, and has a rather amusing slogan: the "#1 for fossilized #2." The coprolite was given the nickname "Barnum" after Barnum Brown, who discovered the first *T. rex*. Clearly, an honor. Like the Canadian coprolite, this second specimen also contained chunks of bone, but it was even larger than its predecessor, measuring 67.5 × 15.7 centimeters. Quite the poop at over 2 feet long! Barnum even entered the Guinness World Records as the largest-known carnivore coprolite.

These Cretaceous coprolites tell us quite a bit about the eating habits of *T. rex*. Most notably, studies of bite force in *T. rex* have consistently shown that it had the most powerful bite of any terrestrial predator ever, capable of smashing through and pulverizing bone. We now know, as predicted,

BOB NICHOLLS ART

that *T. rex* swallowed bones as well. When you have such massive, powerful jaws packed with banana-sized teeth, you would not necessarily be such a fussy eater and would happily crunch through and swallow bone. However, as Chin and her team discovered, the high percentage of incompletely digested bones inside of their coprolite shows that rex did not fully digest bones, unlike hyenas today, for example.

With all this talk of prehistoric animal poo and the fact that there is a specific branch of paleontology specializing in coprolites, it is amusing to perhaps consider and label all these coprolites as "ghost poos". You might have heard of ghost poo before, a phenomenon when you do a number two that leaves no trace. In this case, although it might sound a little odd, coprolites are the poops of prehistoric animals long gone, with only their fecal fossils left behind as evidence that they were ever even here.

FIGURE 9.2. (Overleaf). *His Majesty's Excreta*

Halfway through his meal comprising a *Triceratops prorsus*, this bull *Tyrannosaurus rex* pauses to defecate. It is a dangerous moment for a passing lizard!

THE SCOOP ON SOME RATHER UNUSUAL POOP

Diving into the world of fossil poop can be fun, especially if you love learning about ancient excrement. Beyond giant *Tyrannosaurus* poop, I figured it would be worth opening up your world of ancient jobbies just a little more but add something of a twist. Offering up their secrets, coprolites may preserve traces of an animal's last meal, although sometimes those meals might not be what you expected or could even be a little sinister.

With countless coprolites known from around the globe, ranging from hundreds of millions of years ago to much more recent times, there is an unquantifiable number of ancient ghost poos. Scouring the literature of prehistoric scats, which was as fun as you can imagine, on the hunt for something a bit unusual, I found many coprolites with intriguing stories to tell. Most, naturally, focus on what we might expect—bones or scales being found inside the coprolite of a predator—but some of them challenge our conventional ideas about diets, reveal unique evidence of behaviors, and will no doubt amaze you. Or disgust you.

On that latter point, if you are a little squeamish then you might want to skip the next couple of paragraphs, but know that if you do then you will be missing out. This first example takes us back to around 270 million years ago, to the Middle-Late Permian, and to an ancient lake where a type of early shark had just dropped a poop that came to rest on the lake bed. Luckily for us, the conditions were exactly right to transform the feces into a coprolite, which was found at a site in the municipality of São Gabriel in the state of Rio Grande do Sul, southern Brazil.

This small coprolite, which measures 5 centimeters in length and 2 centimeters in diameter, has a distinct, alternating spiral shape that is typical of elasmobranch (such as sharks and rays) coprolites, a feature and identification that has been recognized since Anning's time. On inspection under the microscope, the coprolite contained scales and bone fragments, typical of what you would expect for a shark, but nestled inside this coprolite was also a cluster of unusual oval-shaped structures. Ingested eggs, maybe? Not quite. These are tapeworm eggs, meaning this shark was infested with parasites.

As per their name, tapeworms are shaped somewhat like a tape measure and are intestinal parasites that can live inside the digestive tract where they feed off the food being digested. The ninety-three tiny eggs have smooth shells and are grouped in small segments measuring just 4 millimeters long and 1 millimeter wide. Several of the eggs appear to have been broken, whereas at least one contains a developing embryophore (the membrane around a mature tapeworm egg) and has both a yolk and shell.

Some tapeworms, or bits of them, along with their eggs, may come out of the body in feces, hence their presence inside this spiral shark coprolite. The shape and structure of the eggs, along with their location in an elongated segment, correlates perfectly with what we know in modern tapeworms. Although they are tapeworm eggs, the researchers ruled them out from being the same species as any living tapeworm as they are larger in size. However, based on some features, they did consider whether they could be members of a group known as Tetraphyllidea, the most widespread group of tapeworms found inside elasmobranchs today.

Many living tapeworms rely on several hosts to complete their life cycle. For example, a small animal eats something contaminated with tapeworms, which is then consumed by something larger and thus continues the next cycle. It is possible that the tapeworms might have originally been present in the fish that this shark ate. In any event, this fossil represents the oldest record of parasite eggs in a vertebrate coprolite. That is certainly some title, though I am not so sure whether that is a title to be proud of, but it does confirm that the interaction between tapeworms and vertebrate hosts goes back at least to the Permian.

Now we move forward in time to the Triassic, about 230 million years ago, and to something just a little less squishy: beetles. Today, beetles are enormously varied and represent the most diverse group of animals known, with more than four hundred thousand species described. In the Triassic, what we know about beetles almost exclusively comes from flattened fossils that limit detailed study. So, when a team led by Martin Qvarnström of Uppsala University, Sweden, examined a partial coprolite collected from a clay pit near the village of Krasiejów in Poland, it was a surprise to discover near-perfect 3D-preserved beetles buried inside.

The team analyzed this coprolite using a huge, powerful machine called a synchrotron, which uses detailed X-rays to peer inside the rock. Announcing their results in 2021, this nondestructive process revealed the coprolite was filled with many minute beetles, with most belonging to one type of beetle with a body length of just 1.4–1.7 millimeters. They were so exceptionally preserved, with many bearing delicate details of the antennae and legs, that their preservation was almost as good as beetles found inside much younger amber. The most common beetles turned out to be a new type, which the team called *Triamyxa coprolithica*, and represented an entirely new family, Triamyxidae, that were assigned to a group of living aquatic or semiaquatic beetles called *Myxophaga*, which are found in and feed on algae.

This is all the more amazing considering that the cylindrical coprolite is a fragment of what would have been a more complete, elongated specimen. As preserved, it is only 17 millimeters long and 21 millimeters in diameter, which showcases the power of synchrotron science to reveal such a wealth of remarkable evidence in such a tiny fragment of an unassuming coprolite.

The amazing, yet amusing thing is that this extinct type of beetle had to wait inside a prehistoric poop for 230 million years before paleontologists could finally recognize that this species even existed, all resulting from the chance finding of a fragment of fossil feces. As a result, this is the first insect species to be described from a vertebrate coprolite. This echoes the discovery of the new species of lizard found inside a specimen of *Microraptor*, which we discussed previously. I think this tells us that if you wish to try and become fossilized, your chances may actually increase if you are eaten by something larger, though I would not recommend this!

Despite being well preserved, the beetles were found in various degrees of disarticulation, with some only represented by fragments. This, combined with the fact that there were no burrows or traces on the coprolite, confirms that the beetles were ingested rather than living in the poop. That raises the question, then, of what was eating these beetles and what produced this prehistoric poop? The team considered the likeliest candidate to have been an animal called *Silesaurus opolensis*, a dinosauriform, which is basically a reptile that looks like a dinosaur but does not quite have enough features to make it a dinosaur. The interpretation was based on previous research led by Qvarnström, in which multiple coprolites with beetle inclusions, all found at the exact location, were matched up with the best candidate from the same locality and same age, *Silesaurus*. The team hypothesized that this

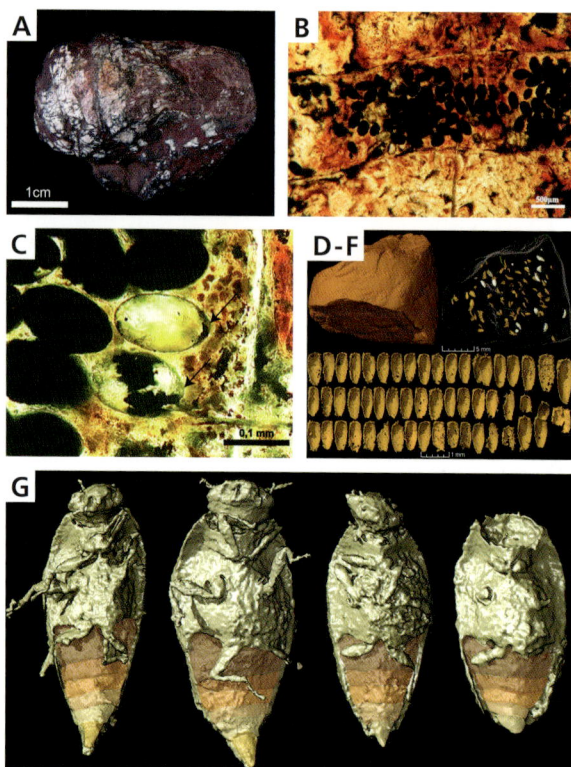

FIGURE 9.3. (A) The shark coprolite containing (B) ninety-three parasite eggs and (C) a close-up of some eggs. (D) 3D rendering of the probable *Silesaurus* coprolite (E) showing numerous visible inclusions. (F) 3D reconstructed *Triamyxa coprolithica* beetles from within the coprolite. (G) Finer details of some of the beetles.

([A–C] Images from P. C. Dentzien-Dias, et al. "Tapeworm Eggs in a 270 Million-Year-Old Shark Coprolite," *PLOS One* 8 (2013): e55007; [D–G] images from M. Qvarnström, et al. "Beetle-Bearing Coprolites Possibly Reveal the Diet of a Late Triassic Dinosauriform," *Royal Society Open Science* 6 [2019]: 181042)

FIGURE 9.4. (Opposite). *Fungus Filled Wood and a Side Order of Crabs*

A *Parasaurolophus cyrtocristatus*, a hadrosaurid dinosaur, hits the jackpot and finds some sodden deadwood covered with fungus *and* a consortium of tasty crabs.

BOB NICHOLLS ART

roughly 2-meter-long animal might have used its beak-like jaws to peck at small insects.

If correct, and the putative identification seems reasonable, then this dinosauriform ate beetles and was partly insectivorous. Given their tiny size, there is also the possibility that it accidentally ingested the beetles while they hung out in algae, the remnants of which are also preserved in the coprolite, when it was foraging for something else, and thus swallowed the cluster of beetles as a by-product of feeding. Still, fragments of larger beetles are present in the coprolite, and some of the other coprolites found previously with beetles inside were also attributed to *Silesaurus*, so this seems to have happened time and again and suggests that insects were most probably a regular food source.

Oh, and reading this quote from Qvarnström in the media, I had to laugh: "We really have to thank *Silesaurus*, who was probably the animal that helped us collect and preserve the beetles." This is a fair acknowledgment considering that without these beetles being swallowed and pooped out, we would never have heard of *Triamyxa coprolithica*. I guess *Silesaurus* was the first entomologist. A real pioneer in the field.

Coming off the back of the beetle-eating dinosauriform, one other recent find that I wanted to mention was some large, up to at least 20-cm coprolites attributed to megaherbivorous dinosaurs. You might question what is so unusual about coprolites made by herbivorous dinosaurs, but multiple coprolites collected from ~75-million-year-old rocks in southern Utah were found to contain a mash of rotten conifer wood and, uniquely, crustaceans!

The research, led by Chin, was published in 2017 and revealed that these paleo poops were produced by dinosaurs that were targeting rotting wooden logs that had decayed due to white-rot fungi. Scattered throughout these wood-filled coprolites were thick fragments of sizable crustaceans (approximately 5 centimeters) that had sheltered inside the rotten logs. The coprolites were collected from at least three distinct levels within a rock unit referred to as the Kaiparowits Formation. The bones of hadrosaurs are most common in this formation and include large, hefty beasties like *Parasaurolophus* and *Gryposaurus*. They are highly likely to have been the dinosaur defecators, especially when considering the crushing and shearing abilities of their dentition, which enabled them to feed on a broader range of foods.

Supplementing their diet with decaying wood and crustaceans would have provided additional nutrients, especially with added animal protein and calcium from the nutritionally rich, crunchy crustaceans. The exact type of crunchy critters remains a bit of a mystery as the remains were not complete enough for identification; however, some crabs have been found in slightly older rock formations nearby, which might suggest they were parts of ancient crabs. This rather surprising evidence challenges our preconceived ideas about large herbivorous dinosaurs solely eating plants. As Chin and her team indicated, this dietary shift was likely to have been seasonal and most probably did not depict the typical, year-round feeding habits of these dinosaurs.

As I am sure you will agree, coprolites present a seemingly untapped world of potential. As technologies continue to evolve and access becomes more available, ancient fossil feces will continue to reveal incredible new insights into prehistoric life. Who knew that an ancient animal taking a number two would become so invaluable to paleontology?

DON'T FORGET TO EAT YOUR STONES

When people ask, "Where did your passion for paleontology come from?" or "How long have you been interested in dinosaurs, rocks, and fossils?" I am very quick to say that my interest goes back as far as I can remember, to perhaps my earliest memories of playing with toy dinosaurs and watching *The Land Before Time*. Technically, you could say that my interests sort of go back a little earlier to something a bit odd—"chewing" on stones as a toddler. This is a behavior known as mouthing, where babies explore by putting objects into their mouths to help them learn and develop. My mum used to tell me that in the school playground, her friends would actively bring me large stones to, uh, "chew on." So, as strange as this is, and let's face it, this is rather odd, I guess you could say my introduction started there. Now, I ponder whether any of those stones contained a fossil. Hmm . . . or should it be mmm?

I recall conversations with my mum telling me that I never swallowed any of these stones, which is a relief, but some animals intentionally swallow stones and may even rely on them partly for their survival. They are called gastroliths or "stomach stones" when found inside the digestive tract. But why exactly would any animal engage in lithophagy, that is, *eating* a stone? Strictly, they do not eat the stone because it does not get digested, but animals swallow them for various reasons, although primarily to aid with digestion, as the stones help to crush and grind food. For instance, if you have chickens, then you will know that grit is used to help them digest their food as it is stored in the gizzard. Crocodiles are a famous example of an animal that swallows stones; as they tear off large chunks from their prey or swallow them whole, the stones may help to grind up the food inside the gut.

We have plenty of prehistoric animals with associated gastroliths despite still being a relatively rare association. Rather than going through them all because that is probably a book in itself, I wanted to include a few rather impressive examples. Besides several birds, such as *Jeholornis*, or birdlike dinosaurs such as the peacock-sized *Caudipteryx* and the bizarre bat wing–like *Ambopteryx*, many dinosaurs have been found to contain a mass of gastroliths. One of the best examples is found in yet another familiar dinosaur, *Psittacosaurus*: the "tacosaur" ceratopsian whom we met previously.

Contained inside one specimen from Mongolia, first studied in the 1940s, was a large collection of 112 multicolored stones within the rib cage. Some paleontologists have noted that gastroliths are, to an extent, found associated with nearly all well-preserved psittacosaur skeletons and that they are typically larger (sort of penny-sized) than what might be expected in a Labrador-sized dinosaur. Considering that there are a lot of specimens with gastroliths, it suggests that they must have been important for this dinosaur. It has been proposed that such a mass of large gastroliths adds further support for the idea that tacosaurs may have predominantly had a tough, high-fiber diet of nuts or seeds that the gastroliths helped pulverize.

Outside of these discoveries, there is quite an extensive list of dinosaurs found with gastroliths recorded from sites around the world. Even though they are still relatively rare, the most recognized group that seems synonymous with gastroliths is the sauropods. As these herbivorous, often giant dinosaurs lacked extensive chewing capabilities, it has long been thought that they must have used gastroliths to aid in digestion. However, there is debate among paleontologists (as ever) regarding how significant gastroliths were for food processing among sauropods, notably because there is an abundance of articulated or partly articulated sauropods worldwide but only a few of these are found with bona fide gastroliths.

Whatever the case, various studies have reported gastroliths with sauropods, including some found with such Jurassic legends as *Diplodocus*. For instance, in the holotype specimen of *Diplodocus hallorum*, once called *Seismosaurus*, from New Mexico, this partial skeleton was found in association with more than 240 stones, many of which were in the rib cage, although again there is debate around whether these are definitely gastroliths. I also mention *Diplodocus* because I fondly remember personally finding what might have been a gastrolith alongside some likely *Diplodocus* bones while on a dig in Wyoming. One definite example that paleontologists seem to agree on was a mass of 115 stones found in the stomach cavity of the brachiosaurid sauropod *Cedarosaurus* from the Early Cretaceous of Utah.

One additional dinosaur discovery that is quickly worth mentioning corresponds to the first unambiguous account of gastroliths in an ornithopod. In 2008, small gastrolith clusters were reported inside the abdominal region of three skeletons of the turkey-sized herbivore *Gasparinisaura*.

In one of the specimens, much of the rib cage was filled with up to at least 180 stones with an average length of 7.9 millimeters. As the authors discussed, this find provides clear evidence that these little ornithopods purposely swallowed the stones, which were likely used to help digest tough vegetation. I also wanted to mention this one because it leads into a published study that I undertook with two friends in 2022, based on the rare discovery of gastroliths found with a much larger ornithopod, *Tenontosaurus*. This was pretty cool as we had been working on this research for several years, and it is genuinely a rare find, representing the second oldest occurrence of gastroliths in an ornithopod and in a more derived, much larger member of the group.

Moving from land and into the water, there are several records of gastroliths in marine animals, most notably in plesiosaurs. These marine reptiles have the best record of any fossil vertebrate with preserved gastroliths, including many skeletons with associated stones. Gastroliths have been recorded in multiple plesiosaur families, from long-necked elasmosaurs to massive-skulled, short-necked pliosaurs.

Of all plesiosaurs, gastroliths have most commonly been found in elasmosaurids, for which an extensive body of literature exists. There are several rather excellent examples, though a couple of highlights, at least for me, include a glorious elasmosaurid called *Albertonectes vanderveldei*, described in 2012 from the Late Cretaceous of Alberta, Canada. With a staggering seventy-five or seventy-six neck vertebrae, this record-setting plesiosaur has the highest number of neck vertebrae known for any animal—living or extinct. It also has at least ninety-seven chert gastroliths present in a tight cluster, with more very likely to have originally been there, but the specimen was discovered when an excavator bucket went right through the fossil. These things happen. The gastroliths are all black or gray in color, are smoothly rounded, and range in size—the smallest is 4 millimeters in diameter, whereas the largest is a whopping 135 millimeters.

Perhaps the most stone-laden plesiosaur was an example of another elasmosaurid, *Aristonectes*, this time from the Late Cretaceous of Seymour Island in the Antarctic Peninsula. This skeleton was found with 793 stones inside. They ranged in size from 4 to 64 millimeters. One of the funny things with gastroliths is that when you weigh them, they weigh the same as when they were swallowed. In this case, the mass of the visible stones was 6.4 kilograms. It is definitely worth mentioning that another elasmosaurid from Antarctica was found with a staggering 2,626 gastroliths! That's a lot of stones, but they only had a combined weight of just 3.02 kilograms, which is interesting considering that the largest stone found in *Albertonectes* weighed 1.13 kilograms.

Further to all of this, it is interesting that plesiosaur skeletons are found in marine deposits that are far away (sometimes hundreds of miles) from any possible source for pebbles like those found as gastroliths. As several authors have surmised, this indicates that plesiosaurs must have gotten rather close to the shore into estuaries and maybe entered rivers on the hunt for these stones. A few plesiosaurs have been found with sand in their abdominal region, too, demonstrating that at least some of the stones were ingested in a beach environment.

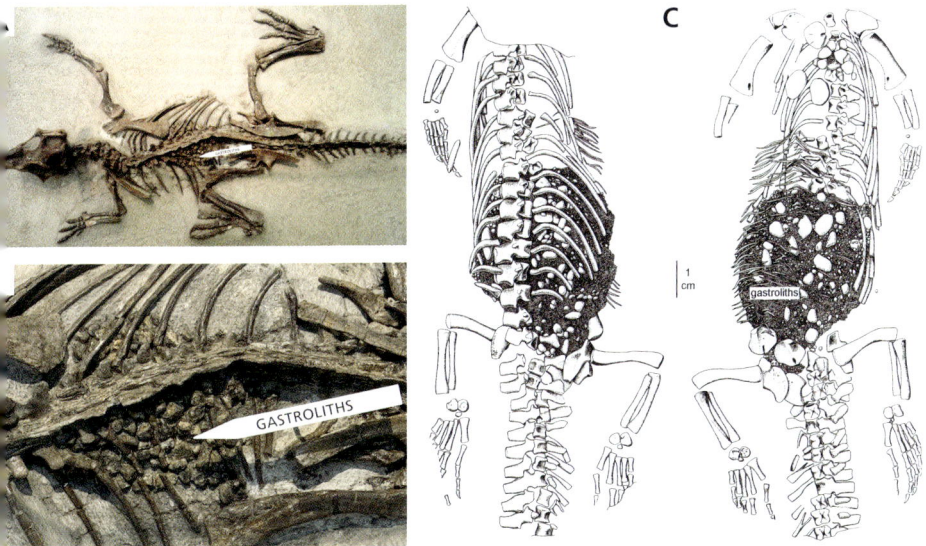

FIGURE 9.5. (A) Gastroliths (stomach stones) in a *Psittacosaurus* skeleton containing (B) 112 multicolored gastroliths. (C) Illustration of a tangasaurid, *Hovasaurus boulei*, with a huge number of gastroliths inside.

([A] Courtesy of Wikimedia Commons; [B] courtesy Bloopityboop, Wikimedia Commons; [C] courtesy of Spencer Lucas)

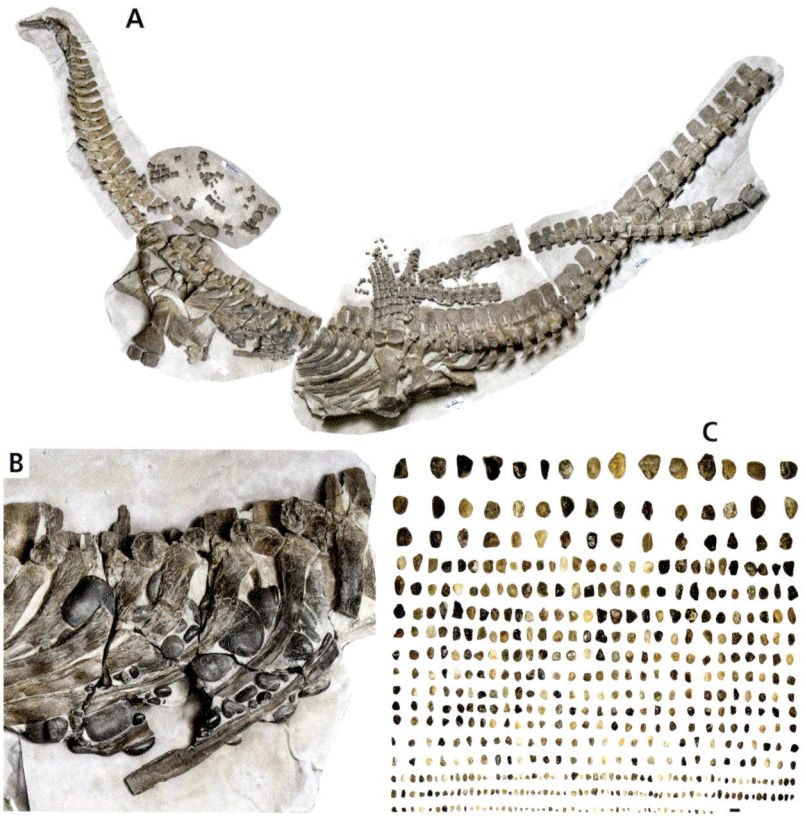

FIGURE 9.6 Plesiosaurs and their stones. (A) The *Albertonectes vanderveldei* skeleton from the Late Cretaceous of Alberta, Canada; tail to the left and very long neck to the right. (B) A close-up of the rib cage showing numerous gastroliths. (C) 534 of the 793 gastroliths found inside the skeleton of *Aristonectes* from the Late Cretaceous of Seymour Island, Antarctic Peninsula.

([A–B] Courtesy of the Royal Tyrrell Museum; [C] courtesy of José Patricio O'Gorman)

FIGURE 9.7 (Opposite). *The Stone Swallower*

Each stone this very long-necked elasmosaur (*Albertonectes vanderveldei*) swallows must travel 8 m along its neck before reaching its stomach!

BOB NICHOLLS ART

Evidently, gastroliths played some sort of role in plesiosaurs, but the exact role is unclear. Their function has been debated for over a century and remains controversial, with the leading theory being that they helped with digestion. Others have argued that the stones might have been used as ballast to help control buoyancy, although this has been challenged, while some have suggested that it may have been a combination of both. In a recent 2024 study by a Royal Tyrrell Museum paleontologist Don Henderson (another fossil friend), his research focused on four Cretaceous plesiosaurs with gastroliths, including *Albertonectes*. His findings revealed that the total mass of these stones was never more than 0.2 percent of total body mass and thus would have been ineffective in helping these plesiosaurs control their buoyancy.

This leads nicely to a creature called *Hovasaurus boulei*, a member of a Permo-Triassic group of aquatic reptiles known as tangasaurids. Lizard-like *Hovasaurus* is the oldest example discussed herein and comes from the Late Permian of Madagascar around 250 million years ago. In 1981, the renowned Canadian paleontologist Phil Currie, famed for his extensive research on dinosaurs and who helped found the Royal Tyrrell Museum in Drumheller, Alberta, among many other things, wrote a comprehensive study on *Hovasaurus*. This included specimens that were jam-packed with pebbles. Some of the stones were even larger than the animal's vertebrae.

He provided a detailed analysis of these pebble masses and their potential function, ultimately leading him to conclude that they were swallowed for ballast rather than digestion. His interpretation inferred that the stones were stored in the abdominal cavity, within a specific sac that was adapted to hold them, and that this shifted the center of gravity to maximize the use of its long tail for propulsion. Estimating the weight of the tangasaurid and the pebbles, Currie showed that it would raise the individual's gravity by about 5–10 percent.

To help with this ballast interpretation for *Hovasaurus* and indeed for plesiosaurs, some help might come from modern-day stone-swallowing crocs. The idea of swallowed stones being used for ballast comes from observations and studies of crocodilians, of which several species are known to swallow stones. Despite the continued back-and-forth research, some recent work has also provided further evidence that crocs use stones for buoyancy control, allowing them to get more oxygen in their lungs and weigh them down, thus enabling them to increase their time underwater. Many modern seals are also known to have gastroliths, so there is certainly some function behind it. Nonetheless, scientists still do not quite see eye to eye with the ballast interpretation, though there are good arguments for it, at least in some cases.

The presence of gastroliths in various dinosaurs and marine reptiles showcases how many different species, often dramatically different—think tiny bird to plesiosaur to sauropod—utilized gastroliths. They must surely have performed some function, likely to aid in digestion overall, but as in the marine realms, it seems plausible that gastroliths may have helped them control ballast, at least in some smaller types like *Hovasaurus*. The discovery of gastroliths inside any prehistoric creature cracks open new ideas about the digestive systems of said creatures, providing further evidence of diet.

I cannot end this without mentioning one of my personal favorites. Here, I loop back to my opening lines about my own somewhat lithophagous behavior and ponder whether any prehistoric animal swallowed stones containing fossils. Imagine finding a half-a-billion-year-old trilobite inside a rock within a dinosaur's stomach from 70 million years ago. That would be a surefire way to mess with a paleontologist's mind. Although nothing quite like this has been found, at least so far, my favorite example close to this is the case of pregnant gopher tortoises in South Florida that swallow calcium-rich fossil seashells and use them as gastroliths. Somehow, this feels like a strange circle of life moment.

COUGH IT UP!

I'm sitting having my lunch and the dog is pestering me for food. Lucie, my little Yorkshire Terrier, will not give up hope of getting even the tiniest morsel. First, it's cute little murmurs of noise, then frantic running, yapping, and barking, and then she appears on my shoulder like a parrot. Finally, I reluctantly give in, save her a little piece, and make her sit and give me her paw. She gets a little treat. A couple of minutes later, I hear that oh-so-distinct retching noise. I jump up and run into the kitchen, and there's the now mashed-up bit of undigested food sitting on the floor and the dog staring lovingly at me. If you can relate, I feel you.

Of course, this doesn't just happen in dogs. Regurgitation of food is a common behavior across the animal kingdom that occurs for several different reasons. Humans might involuntarily regurgitate food, perhaps because it irritated us on the way down, we ate too fast, overindulged, or it could be due to acid reflux. Yes, you might think it gross, but it is a vital activity in the lives of many animals.

Birds of prey such as owls are famously known to intentionally regurgitate as a post-meal-ritual. They cough up a compacted mass called a gastric pellet, comprised of indigestible material formed by compaction in the gizzard, which may consist of skulls, teeth, claws, fur, and feathers. Dissecting an owl pellet can provide clear clues about what the bird has been eating.

In the fossil record we have evidence of regurgitation in the form of prehistoric gastric pellets, offering yet another opportunity to directly examine the last meal of an animal beyond coprolites or stomach contents. These are trace fossils typically formed of a compact skeletal mass that have a special name and meaning. Called regurgitalites, they are evidence of any digested or partially digested food material egested (removed from the body) via the oral cavity. Some types are even further defined, such as emetolites, which are from an animal that habitually egests pellets.

A fair haul of prehistoric pellets has been reported, some more speculative and suspect than others, but some are much more convincing and represent genuine pellets rather than just an oddball mash of bones. As such, I figured discussing a couple of my favorites would be appropriate. One of these includes a return to a familiar locality discussed previously—Egg

Mountain in northern Montana. We know from fossil evidence that mouse-sized mammals were living in the same area as hefty herbivorous dinosaurs at this location. These mini mammals, however, did not have it easy.

Two accumulations collected at the site were found to contain three and eleven individuals, respectively, of mashed-up mammals. Besides one containing a lizard, all the remains consisted of *Alphadon halleyi*, a relative of modern marsupials. Chunks of skull, broken jaws, several teeth, and other indigestible parts, combined with broken and disarticulated elements and evidence of acid corrosion, identify both fossils as regurgitalites.

As for who was snacking on these mini mammals, an abundance of shed teeth and nesting evidence, including eggs, at the Egg Mountain locality suggests that the culprit may have been turkey-sized juveniles of the bird-like theropod *Troodon formosus*. This dinosaur had large eyes and excellent vision. This led some researchers to propose that *Troodon* might have been nocturnal or crepuscular, so perhaps it hunted these mammals under the cover of darkness. Amazingly, these two specimens represent the oldest known mammal-bearing regurgitalites in the world.

Identifying the predator that left behind its coughed-up regurgitate is hard and often speculative, based on circumstantial evidence, but not always. There are some stunning cases where we know precisely who the predator was. In a sideways step from dinosaurs for a moment, a pair of pterosaurs, a juvenile and an adult, from the Late Jurassic of Linglongta in Liaoning China were each found with an associated gastric pellet. This is not the first pterosaur regurgitalite to be reported, but it is the only conclusive evidence so far.

Both pterosaurs are virtually complete and belong to a small, crow-sized species called *Kunpengopterus sinensis*. The pellet is immediately adjacent to each pterosaur, located just outside of the body, suggesting that it was expelled either shortly before death or as part of the decay process. Each pellet represents a small, subcircular dark mass containing densely packed fish scales. The scales belong to the same type of fish and suggest that the juvenile ate small individuals, whereas the adult dined on larger individuals. In both cases, the associated pellets were identified as emetolites and represent the only convincing evidence of pterosaur gastric pellets so far.

To a point, some skeptics might say that the pterosaur-pellet association in both cases was purely a random coming together. I highly doubt that for the reasons above, but to quash any remaining doubts about regurgitalites, look no further than the next extraordinary fossil—the showpiece.

Sticking in the Late Jurassic of Linglongta China, we return to some remarkable, exceptionally preserved dinosaurs. In 2018, six gastric pellets were found to be attributable to a cousin of the aforementioned *Troodon*, called *Anchiornis*. Where this pigeon-sized Chinese cousin sits on the dino-bird family tree is a bit contentious, but it is a member of the group known as Paraves, which contains true birds and all their closest relatives, such as troodontids and the famed dromaeosaurids (or "raptor dinosaurs"). Some paleontologists think it is an actual troodontid.

In a similar scenario as the pterosaurs, three *Anchiornis* skeletons were each found directly associated with a single, highly compact, oval structure containing a compact mass of fish scales and some fish bones. Likewise, two other specimens were found in isolation but bore the same shape as the other structures, comprised of fish scales, and were collected from the same rock layers of the *Anchiornis* remains. All five oval structures represent pellets.

But what about the sixth specimen? This *Anchiornis* skeleton is surrounded by a layer of fuzzy feathers and soft tissues, and a pellet is perfectly lodged in the esophagus. This oval structure measures just 7 centimeters long and immediately occupies the region in the throat posterior to the back of the beautifully preserved jaws. Contained inside the pellet are the partly disarticulated bones from at least three individual lizards that this theropod must have eaten in swift succession; the lizards are surrounded by fine-grained white sediment that resulted from digestive residues. The position of the pellet suggests that this *Anchiornis* was readying to regurgitate before death. It is highly unlikely that it choked to death on the pellet.

All these *Anchiornis* pellets are remarkably comparable to those produced by modern birds. In each pellet, the bones and scales have relatively smooth surfaces and show only minimal evidence of light acid etching, implying that the remains were exposed to gastric juices for just a short period of time just like in regurgitating predatory birds. Again, like modern species, these *Anchiornis* probably swallowed their lizard and fish

FIGURE 9.8. (A) Fresh owl pellets containing undigested bones, including part of a vole skull. (B) Parts of a paleo pellet containing chunks of jaw and broken bones of the marsupial relative *Alphadon halleyi* from Egg Mountain, Montana. (C) The adult *Kunpengopterus sinensis* pterosaur with associated pellet containing (D) the remains of fish.

([A] Courtesy of Breanna Martinico; [B] courtesy of William Freimuth and David Varricchio; [C–D] courtesy of Shunxing Jiang and Wei Gao)

FIGURE 9.9. (A) The exceptionally preserved *Anchiornis* with a pellet perfectly lodged in the throat. (B) A close-up of the pellet containing numerous bones from at least three lizards.

(Images from X. Zheng, et al., "Exceptional Dinosaur Fossils Reveal Early Origin of Avian-Style Digestion," *Scientific Reports* 8 [2018]: 14217)

Bob Nicholls Art

prey whole or dismembered them into large, edible chunks. These pellets are the only examples confidently known to have been produced by a nonavian theropod.

Confirmation that *Anchiornis* could produce and egest gastric pellets is significant. The discovery reveals that an avian-style digestive system was already present over 160 million years ago among one of the closest cousins of birds. Intriguingly, as we discovered with *Microraptor* in a previous chapter, the evidence from that little dromaeosaurid shows that it swallowed prey whole and readily ingested the bones. Something similar can be said for other theropods where we have direct evidence, such as the whopping big *T. rex* turd with bones inside, again confirming as in *Microraptor* that the hard parts were expelled through the cloaca rather than regurgitating them. This all suggests that *Anchiornis* could digest food faster and more efficiently like birds than other dinosaurs, at least based on our current knowledge.

These prehistoric pellets offer us a little peek into predator-prey food webs and the function of the digestive system. This, therefore, provides further evidence of the lifestyle and physiology of extinct species than what we might expect to learn from stomach contents or poop. We always knew that said animal would have eaten and pooped, but we never would have known that it regurgitated its food as a way of life, thus informing us about the evolution of dietary processes.

Identifying the types and numbers of critters the animals ate, and what species, along with evidence of damage and corrosion by digestion, are the same telltale signs that we look for when dissecting modern gastric pellets, like those of an owl. This is just one of the many cool, exciting ways that rare fossils like these help us to connect the present with the very distant past. Who knows what the fossil record might throw up next?

FIGURE 9.10. (Opposite). *Up It Comes!*

It took some effort, but at last this *Anchiornis huxleyi* (a birdlike theropod dinosaur) coughs up a gastric pellet containing undigested lizard bones.

10 | Health and Endgame

A pack of feral dogs chased a young pooch into the crocodile-infested Savitri River in Maharashtra, India. Onlookers watched, expecting the three circling crocs to move in on the easy target when something unexpected happened. Two of the three crocs came to the dog's rescue, nudging him to safety on the riverbank away from the pack of dogs. A possible case of cross-species empathy. It is not all doom and gloom.

DINOSORE

Just like us humans and other vertebrates, dinosaurs got injured and dealt with broken bones from time to time. We have a rich record of dinosaur bones that were broken in life, with pathological specimens showing that they then sometimes lived a rather difficult, troublesome existence. Of course, we should not think of dinosaurs as unfortunate oddballs that routinely broke bones on a daily basis—by clumsily tripping over and breaking a leg, falling down holes, or getting into one too many fights. Although these things certainly happened on occasion, and could genuinely be life-threatening at times, dinosaurs, just like modern vertebrates, would have broken bones for a wealth of different reasons.

Reading the word *pathology* might evoke a sense of nostalgia for some of you. Not for those of you who have ever broken a bone (me included) but for those who might recall watching "The Ballad of Big Al," a 2000 spinoff special episode of the rather famous BBC series *Walking with Dinosaurs*. Much like how *Jurassic Park* transformed our view of dinosaurs in Hollywood, *WWD* was the first mega-dinosaur series that blew audiences away with its CGI and storytelling; as a ten-year-old, I remember being enthralled by this series, and I watched it on repeat. Set in the Late Jurassic, "The Ballad of Big Al" followed the life of a single, subadult *Allosaurus*, nicknamed "Big Al," from hatching out of his egg to his final curtain call. It introduced the viewer to dinosaur pathologies and showed dinosaurs as real-life creatures.

A broken toe here, a cracked rib there, Big Al had it rough. The fact is most of what you watched and learned about Big Al was real. It was based on a tremendous specimen of *Allosaurus* found in 1991 at the Howe Ranch Quarry near Shell, Wyoming, which revealed almost twenty injuries, including an infamously broken toe that became infected. This evidence of behavior is what helped connect people with his character and story to the point where you could genuinely feel sad for this prehistoric predator. Although he overcame several of his injuries, unfortunately for Big Al his ballad would be cut short; he lived fast and died young, probably aged thirteen to fifteen years old. The fast-paced life caught up with Big Al as his battered body forced him to walk with a limp and fight various infections, finally succumbing to what was probably a painful death before ever reaching adulthood.

Since Big Al's initial discovery, there had been a buzz about this exceptional *Allosaurus*, notably because of its great completeness, the pathologies it displayed, and the fact that it was probably a new species. Fast forward twenty years from its big TV doc and Big Al was finally described in detail and named as a new species, *Allosaurus jimmadseni*. Much of Big Al's remains are on display at the Museum of the Rockies in Bozeman, Montana, but additional specimens of this species are also known, like "Big Al 2," which curiously enough also has various pathologies. Examples of other *Allosaurus* specimens have been found with numerous pathologies, too, showing that this large theropod lived dangerously.

Theropods appear to reveal the most evidence of injuries among dinosaurs. This is likely due to an active and somewhat precarious predatory lifestyle, as embodied by Big Al. This makes sense considering that predatory theropods, like animals on the hunt today (think lions attacking wildebeest), are more prone to getting wounded—sometimes fatally—during predator-prey interactions.

Stalking and attempting to take down your meal can be physically exhausting, and your meal can also fight back, whereas herbivores do not need to defend themselves from plants that are trying to eat them. Sure, predators will hunt them, but this doesn't happen every single day. Besides hunting for food and taking on the ramifications of what might have come with that, theropods would have had to compete with other rival theropods, sometimes larger species, and will no doubt have fought with conspecifics over territory, mates, and carcasses. Thus, life as a theropod was challenging, and a broken bone might have become debilitating or even signify the end in some circumstances.

One case in point: A study published in 2016 revealed a record-breaking number of bone abnormalities in the forelimb and pectoral girdle of a theropod dinosaur. By coincidence, that dinosaur was yet another species made popular with the public by the media, the crested *Dilophosaurus wetherilli*. In *Jurassic Park*, some next-level artistic license depicted *Dilophosaurus* with a neck frill and a supposed ability to spit acid, two features that *Dilophosaurus* regrettably lacked. It is also seen in the movie as a small

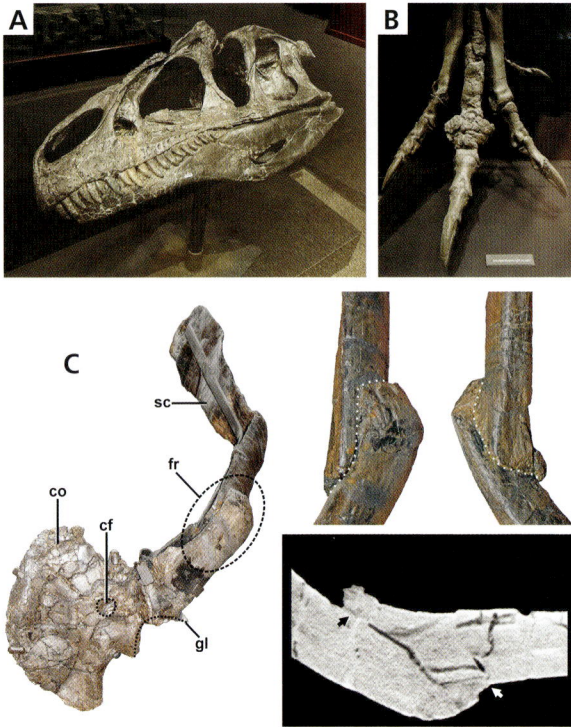

FIGURE 10.1. (A) A close-up of the skull of Big Al on display at the Museum of the Rockies, Montana. (B) Big Al's right foot showing the broken, infected toe bone. (C) The broken and reset left scapula of Big Al 2.

([A-B] Photographs by the author; [C] image modified from C. Foth, et al., "New Insights into the Lifestyle of *Allosaurus* [Dinosauria: Theropoda] Based on Another Specimen with Multiple Pathologies," *Peer J* 3 [2015]: e940)

FIGURE 10.2. (Opposite). *The Battle-Scarred Warrior*

Life as a top predator can be hard. This *Dilophosaurus wetherilli*, a theropod dinosaur, has multiple injuries caused by prey that fought back and other rival *Dilophosaurus*. The gaunt, emaciated body reveals that this individual has been struggling to find food consistently. Only time will tell if it will survive.

BOB NICHOLLS ART

theropod that does not like to play fetch. In contrast, the reality is that it was a medium to large predator, reaching about 5–7 meters in length, and probably enjoyed fetching small to medium-sized prey, including other dinosaurs and the occasional Dennis Nedry in *Jurassic Park*.

The bruised and battered *Dilophosaurus* in question comes from ~193-million-year-old, Early Jurassic rocks of northern Arizona, and, rather interestingly, is the holotype of the species. That is, the founding specimen of *D. wetherilli* that was first described over seventy years ago based on a partial skull and incomplete skeleton. Significantly, *Dilophosaurus* was the largest theropod known in North America during the Early Jurassic and would have played an important ecological role. As the top predator in its environment, just like *Allosaurus*, it would have been subjected to similar dealings with its prey and competition through intraspecific interactions. And, again like Big Al, there would always be a chance that this could lead to injuries and other ailments.

Among theropods, the pectoral girdle and forelimb are frequently found with some injury or other condition. A reexamination of this region in the *Dilophosaurus* holotype revealed four afflicted bones on its left side, including a fractured scapula and radius and abnormal bony growth in the ulna and thumb phalanx. The right side also displayed four oddities: an unusually twisted humerus shaft, bony tumors on the radius, and two deformities in the third finger. Up until this study, no dinosaur was known with more than four afflicted bones of the pectoral girdle and forelimb, whereas this *Dilophosaurus* had eight. A real pain in the . . . arms.

Affected bones may sometimes show clear and obvious signs of healing, such as the presence of a bony callus, suggesting that said animal did not die due to its complications, at least not immediately. In this *Dilophosaurus*, evidence for healing and remodeling is observed in each of the eight maladies, showing that the individual survived for a long time after it sustained its injuries, possibly even years. However, the right third finger was permanently deformed, fixed in a set position, and could not bend. It would have made grasping prey much harder. The researchers hypothesized that the unusual nature of the right humerus could be due to excessive use of the right forelimb to avoid pain in the wounded and infected left forelimb. This would have created an imbalance, a noticeable shift in walking gait and an oddly angled right forelimb.

Looking at today's birds and their close cousins, the crocs, many species make for particularly hardy animals and may recover from all sorts of injuries. For instance, I cannot count the number of plucky pigeons I have seen with missing toes or the occasional missing leg. The same goes for crocs that have been documented to live seemingly normal lives despite missing limbs or even severed jaws. If they can adapt and successfully feed, then there is always a chance of survival.

With all that said, for *Dilophosaurus* and *Allosaurus*, like birds and crocs, having severe trauma would render these individuals at risk of further injuries and prone to potentially fatal diseases, regardless of whether it caused minor or major damage. Depending on the severity of the condition, and whether it left the theropod with short-term or long-lasting side effects, it could hamper their ability to catch prey and leave them incapacitated with an increased risk of death.

In the cases outlined here, although it is hard to say if some of these peculiarities were created during dino bouts or from trips or falls, these theropods lived for some time with their conditions, which tells us that they were inherently tough and capable of enduring unbearable pain for extended periods. It might even be plausible that support from members of a gregarious group aided the longer-term survival of these beat-up theropods, helping them live to see another day. Whatever the case, it was a hard-knock life.

TRAUMATIC AMPUTATION

Not all reptiles have the awesome ability to regenerate their tails like our little lizard friends described in chapter 5. You might not think it is the coolest superpower, but the regrowing or healing of a tail could be the difference between life and death. This superpower even made it into the Marvel Cinematic Universe as the "Lizard" character was able to regrow his torn-off tail and survive. So there is that.

We know that some other reptiles, such as crocs, have powerful immune systems and impressive healing powers capable of swiftly healing wounds. Depending on the severity of their injuries, they can live with parts of or even the entire tail missing, with rare instances of regeneration. Many non-fatal tail amputations have been recorded among living crocodilians. But what about dinosaurs? Living dinosaurs—birds—do not have bony tails, of course, but they do have long tail feathers that play essential roles in flight and display, among other things. While it is unfeasible to compare a feathery tail with a bony tail, since they are different structures, birds can indeed survive without their tail feathers. The loss, however, can greatly impact their aerial abilities, force a change in lifestyle, and increase the risk of predation.

To our knowledge, all nonavian dinosaurs had tails, from the exceedingly long, whiplike tail of *Diplodocus* to the unusual finlike tail of *Spinosaurus*. The tail played several important roles among the vast array of dinosaurs: a counterbalance, an offensive or defensive weapon, a sexy display structure, and a stabilizer when running. Given the multipurpose role that the tail served, it is no surprise that dinosaur tails sometimes bear evidence of injury or damage sustained in life.

In 2013, a strange paleopathology was reported in the tail of an Early Jurassic basal sauropodomorph, an early relative of later sauropods. These types of dinosaurs are sometimes informally referred to as "prosauropods." The tail in question probably belonged to one of the most famous of these early dinosaurs, *Massospondylus*, originally described in 1854, and represents one of the first dinosaurs to be formally named.

The fossil comprised a single partial skeleton collected in March 2008 as part of an expedition to a famous fossil formation (called the Elliot

Formation) near the town of Barkly East in Eastern Cape Province, South Africa. The skeleton included part of an articulated spine, including dorsal, sacral, and caudal vertebrae, along with bones from the pelvis and hind limbs. Based on the shape and structure of these bones, the remains were found to match closely with *Massospondylus*, which happens to be the most common sauropodomorph found in this fossil formation.

Following collection, the block containing the tail was prioritized for cleaning and preparation, which revealed twenty-five vertebrae. Rather bizarrely, the last three vertebrae in the sequence were fused together and clearly deformed. The three vertebrae are amalgamated into an overgrown, jumbled mass of what looks like a big triangular block of reactive bone, resulting from the formation of new bone in response to some sort of trauma, but in this case, representing a massive hyperostosis (too much bony growth). Only the curved neural arch of vertebra 25 is present, but it is mangled inside this bony mass. The neural arch seems to have been sub-stantially deformed, with the posterior part appearing to have been sheared off but subsequently fused with the rest of the bone and strongly offset.

This is where things become even more peculiar. Before this set of three pathological vertebrae, the rest of the tail was found in perfect articulation and with everything in place, even including the bones that hang below from the centra (the chevrons). Yet no more vertebrae were found beyond the pathological mass. The tail ends right there. What gives?

There are typically forty-five to fifty tail vertebrae for basal sauropodo-morphs, indicating twenty to twenty-five vertebrae are missing in this *Massospondylus*. That means about half of the total tail vertebrae are absent, with approximately a third of the length of the tail gone. The abrupt end of the tail signified by the big mass of bony regrowth suggests that the tail was somehow truncated in life.

As we know, nonfatal tail amputations are well-known among croco-dilians. For example, a study in the 1960s surveyed tail amputations in the Nile crocodile and found that 89 of 548 individuals (16 percent) had some type of tail injury that often involved the loss of the entire tail end. More-over, injuries to the tail were more common than injuries sustained else-where on the body. It was proposed that these attacks were due to combat between potential cannibals. A suite of other studies focused on different

FIGURE 10.3. (A) Female saltwater crocodile missing most of the end of her tail, at Singapore's Sungei Buloh Nature Reserve. (B) Distal tail vertebrae of the pathological *Massospondylus*, including a close-up of the three fused vertebrae, which form a triangular-like block of bone.

([A] Courtesy of Joxean Koret; [B] courtesy of Richard Butler)

FIGURE 10.4. (Opposite). *The Tail Taker*

After a lengthy battle, the exhausted *Massospondylus carinatus* (a sauropodomorph dinosaur) fell to the ground. The assailant, a full-grown *Dracovenator regenti* (a theropod dinosaur), refused to let go of the tail and after repeatedly biting and shaking, eventually took the tail clean off! A moment later, while the predator is distracted, the *Massospondylus* will make her escape.

croc species have also found that the tail seemed to be the most targeted area. For something a bit different, there are records of tail or ear amputations in black rhinos that are the result of failed predation attempts on young rhinos by lions and spotted hyenas.

The tale of the tailless *Massospondylus* could be interpreted in two ways. The tail end was traumatically amputated, ripped off by a predator. Or, one of the following scenarios may have led to an unusual pathology and the eventual loss of the tail: a previous, unsuccessful predation attempt, a direct trauma (like the tail being trampled), or a devastating infection; perhaps it was a combination of these. The first scenario seems most plausible given that tail vertebra number 25 was sheared in two, thus rendering this an unsuccessful predation attempt. That said, this would mean the predator did come away with part of the tail as a consolation prize, so it was kind of successful.

So, whodunit? Before attempting to answer that, it is worth noting that based on the preserved bones of this *Massospondylus*, it can be estimated that this individual would have had a body length of 6 meters (with the tail) and represents one of the largest examples known, clearly an adult. So, the paleo perpetrator had to be large enough to attempt to take down an adult *Massospondylus*. A few theropods are known from the Elliot Formation and have been found in the same rocks as *Massospondylus*, with the prime candidate being the theropod *Dracovenator*, which might have reached comparable lengths and would have been one of the top predators in the ecosystem.

It is impossible to confirm whether *Dracovenator* really did the damage, considering that the bitten tail was taken away, but it remains a strong probability. As excruciatingly painful and shocking as it must have been for this *Massospondylus*, the bony growth shows that this individual survived the horrific experience and that the injury was not immediately fatal. Nevertheless, chances are that it led to the dinosaur's eventual downfall.

DEADLY DECAPITATION

December 10, 1823, was a cold, wet, wintery Wednesday, the sort of day that the paleontologist and fossil hunter extraordinaire Mary Anning knew so well. While on the hunt along the beach at Lyme Regis, England, twenty-four-year-old Anning found the world's first complete plesiosaur skeleton. This certainly would not be her last, but this discovery was truly ground-breaking and would help fuel a scientific revolution.

Showing her skills as a competent scientist, upon discovering and cleaning the fossil, she compared it with the illustrations of the first fragmentary plesiosaur bones published just a couple of years earlier. She even sketched the skeleton and penned a letter explaining her identification: "One thing I may venture to assure you it is the first and only one discovered in Europe." Putting two and two together, she realized her find was unique, representing the first true view of a plesiosaur. A world first.

At the time, not everybody was convinced this fossil was real. "Fake!" some people exclaimed, with even the leading comparative anatomist, Georges Cuvier, questioning the fossil's authenticity and, in effect, Anning's credibility. After some back-and-forthing and a formal presentation by geologists at a meeting of the Geological Society of London on February 20, 1824, it was determined to be no fake but the real OG *Plesiosaurus*. The specimen is still the most complete example of the genus *Plesiosaurus*, a name entrenched in paleontological lore. Regarded widely at the time as Anning's greatest ever find, and some might say the same today, one of the main reasons it grabbed so much attention was that this marine creature had such an absurdly long neck.

In the two hundred years since Mary's famous find, a substantial body of research has been dedicated to understanding why many plesiosaurs, like *Plesiosaurus*, had such incredibly long necks. For instance, just cast your mind back to chapter 9 to the stone-guzzling *Albertonectes*, which has the most neck vertebrae of any animal ever. The neck makes up almost two-thirds of its entire body length. But could such a long neck also be a prime target for a predator? Even going back to Anning's time, countless reconstructions, such as the famous *Duria Antiquior* (meaning "a more ancient Dorset") painted in 1830 by her geologist friend Henry De la Beche, and inspired by her finds, have vividly shown long-necked plesiosaurs with their necks being munched, crunched, and bitten off. Yet, although an exceedingly long neck might seem like an easy target for big, toothy predators, there is still zero evidence for this type of behavioral interaction in plesiosaurs. Our story does not end here. The cool thing about the fossil record is that we can look elsewhere.

Some other groups of oddball aquatic reptiles evolved inherently long necks, too, with the Triassic *Tanystropheus* representing the ultimate weirdo of them all. That is, of course, a great compliment to the bizarre nature of this unusual reptile that lived during the Middle to Late Triassic. It is one of those creatures that most fossil fans have probably heard about, since it is often featured in books due to the extreme gangly neck but also because it once had a role in part of the *Walking with Dinosaurs* spinoff series *Sea Monsters*. Exemplifying its fame further, there was even a toy made for the original *Jurassic Park* franchise that I remember owning as a kid.

Most *Tanystropheus* fossils come from an exceptional fossil locality at Monte San Giorgio, a UNESCO World Heritage Site on the Swiss-Italian border. The rocks here are 242 million years old from the Middle Triassic and are of marine origin, principally representing a nearshore, shallow marine ecosystem. Many specimens including complete skeletons have been found, providing a comprehensive view of this early reptile. Thirteen very elongated, narrow vertebrae made up the comically long and stiffened neck that was almost three times the length of its torso.

Over the years, scientists have speculated about its lifestyle based on the unusual proportions, namely focusing on the neck, which has been quite a contentious topic. Opinions are divided on whether it was strictly an aquatic animal or whether it lived along shorelines and spent some time swimming in the shallows, fishing with its long, rodlike neck. Whatever the case, paleontologists can agree that it spent at least some time in the water.

In a recent study published in 2023, the paleontologists Stephan Spiekman and Eudald Mujal from Stuttgart's State Museum of Natural History in Germany investigated a pair of *Tanystropheus* fossils from Monte San Giorgio. One of these belonged to the larger species, *Tanystropheus hydroides*, which could be up to 5 meters long, and the other was a species called

T. longobardicus, which reached a little over a meter or so in length. Spiekman and Mujal named the former species a few years before in honor of the uncanny resemblance to the mythical Hydra of Greek mythology.

Both specimens consist of a complete skull attached to an incomplete but otherwise perfectly preserved neck. Finding isolated skulls, neck vertebrae, or fragmentary skeletons is not uncommon. Often, as with many fossil vertebrates, especially large individuals, a specimen might become disarticulated and scattered pre- or post fossilization. In both examples, the slender, articulated neck abruptly terminates at a clear break in a vertebra and its associated, long and delicate ribs. The reason? This pair of *Tanystropheus* were decapitated. Did that make you rub your neck?

Confirming the severed nature, the *T. hydroides* preserves bite traces in the form of two perfectly aligned puncture marks on the tenth vertebra at precisely the spot where the break in the neck occurs. This decapitating bite created a major fracture in the vertebra, forming a large, V-shaped splinter of bone reminiscent of a broken chicken bone. Based on the position of the punctures, along with the bony splinter, it reveals that the predator likely seized the neck from above and from the rear, pulling back and biting at least twice with enough force to swiftly slice the neck in one go.

The altercation was just as grisly in the *T. longobardicus*. This time, a similar break was found in the seventh vertebra, which also had a bone splinter. Interestingly, the fifth vertebra displayed a tooth-shaped oval pit and associated damaged ribs, suggesting that this part of the neck was also bitten. This probably indicates the initial attack came from above, as in the other individual. The smooth surfaces of the fractured vertebrae are typical of that observed in modern fractures of fresh bones at the time of or shortly after death.

In both instances, the predator landed a direct decapitating bite in the midsection of the neck. This was probably a learned behavior on the predator's part, preferentially targeting the prey's weakest link. Severing this section of the neck and completing a successful hunt, the predator presumably then tucked into the meaty, nutritiously dense torso that is conspicuously missing in both specimens.

The question is, who was the predator with a macabre knack for beheading its prey? As the researchers discussed, many predators have been

FIGURE 10.5. The 1830 *Duria Antiquior* painting by Henry De la Beche was inspired by many of Mary Anning's original finds. Note the ichthyosaur biting the neck of the plesiosaur.

(Courtesy of Wikimedia Commons. The original is in the National Museum of Wales, Cardiff)

described from the Monte San Giorgio site, including predatory fish and marine reptiles. Based on the small size of *T. longobardicus*, the hunter could have been any number of large fish or small to medium marine reptiles. For the larger individual, however, it had to have been a large marine reptile with pointed dentition and enough bite force capable of delivering a decapitating chomp.

In an attempt to decipher the killer, Spiekman and Eudald measured the distance between the two punctures on the vertebrae, which came out at 14.5 millimeters. Comparing this distance with the tooth spacing known for large contemporary marine reptiles, they narrowed it down to three suspects: the small-headed, long-armed *Helveticosaurus zollingeri*, the pre-plesiosaur mimic, *Nothosaurus giganteus*, and the ichthyosaur *Cymbospondylus buchseri* (or maybe that should be "butcheri"?). Without seeing

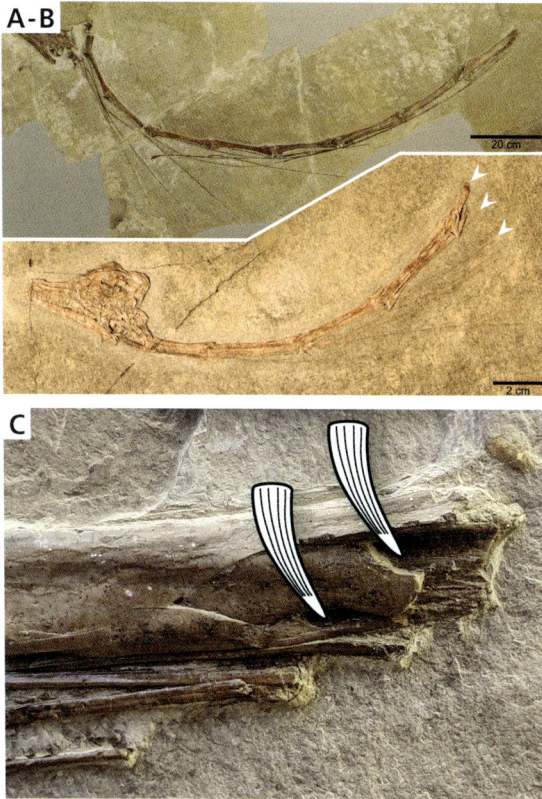

FIGURE 10.6. The pair of decapitated *Tanystropheus*. (A) Head and partial neck of the larger species, *T. hydroides* and (B) the same for the smaller species, *T. longobardicus*; arrows point to fractured vertebra and ribs, which are broken in the same spot. (C) A close-up of the broken vertebra in the larger individual, with a V-shaped bone splinter; teeth added to show the spot of the slicing bite.

(Courtesy of Eudald Mujal and Stephan Spiekman)

them in a police lineup, it is hard to pick the most obvious choice, but most conceivably the similarly sized 4–6-meter *Nothosaurus* or *Cymbospondylus* were the likeliest suspects.

What might all of this tell us about plesiosaurs? Well, considering that we have yet to find a decapitated plesiosaur does not mean that plesiosaurs avoided getting their heads chopped off. Some probably did. What it

Bob Nicholls Art

means, however, is that by contrast with *Tanystropheus*, plesiosaurs were fully adapted to the marine realm and lived comfortably in the water. It was their home. For *Tanystropheus*, it was not a fast or efficient swimmer, and the water was not entirely its natural habitat. Could an exceedingly long, stiffened neck and a lack of aquatic adaptations render *Tanystropheus* vulnerable to attack from a predator? Quite probably.

It is not unreasonable to imagine a *Tanystropheus* paddling along in the water, or perhaps walking along the seabed, using its long neck to search for fish. Then, suddenly, *wham!* Out of nowhere, an ichthyosaur shoots through the shallows and decapitates the *Tanystropheus*. We see modern sharks and whales, including orcas, coming close to the water's edge when hunting prey, so why not ichthyosaurs or indeed nothosaurs? Alternatively, maybe an unsuspecting *Tanystropheus* on the rocks casts out its long neck across the water, biding its time to ambush its fish and squid prey, when an ichthyosaur darts out like a dolphin from the depths and seizes its neck, slicing its head from its body. Although romanticized (and kind of gruesome), for sure, this sort of scenario is not inconceivable.

Many unknowns surround *Tanystropheus* and its lifestyle, making it an appealing animal to study. We will never know for sure what predators were responsible for the deadly decapitations. Chances are, these *Tanystropheus* probably didn't see the attack coming. Having two beheaded individuals suggests this reptile must have been doing something that left itself exposed to risky situations, allowing predators to get close enough to target its weak spot and deliver a fatal strike. If only *T. hydroides* could have summoned its inner hydra abilities and regenerated two heads. Maybe that could have given it a fighting chance.

FIGURE 10.7. (Overleaf). *The Executioner's Meal*

A community of nothosaurs (*Nothosaurus giganteus*) have learned that the quickest way to dispatch a *Tanystropheus hydroides* is to take off its head. The long neck is the target, and once it has been severed the *Nothosaurus* can feed at its leisure.

TWISTED REPTILE

Scoliosis is a condition defined as an abnormal sideways curve of the spine. It can affect people of any age, but it is usually diagnosed during adolescence when growth spurts occur. In the United States, about 2–3 percent of the population, between six to nine million people, are affected by scoliosis, and a cause is usually not found. Most cases are so mild that the condition can go entirely undetected. Extreme cases, however, can lead to uneven posture and pain, resulting in other complications associated with muscle weakness, organ function, and imbalances. Sorry if this sounds like some kind of doctor's appointment (though in reality, it sort of is), but it will make sense.

We might consider scoliosis a very human condition, but it is found in many vertebrate animals, from cats and dogs to fish, birds, and whales. It also affected prehistoric animals hundreds of millions of years ago. There are different types of this spinal condition, too, including congenital scoliosis, which we will focus on here.

This type of scoliosis in humans is thought to affect approximately one in one thousand newborns and refers to a spinal deformity that was present at birth when the spine failed to form during fetal development. One type of congenital scoliosis is caused by the incomplete formation of one or more vertebrae, leading to the development of just one half of a vertebra that acts like a wedge in the spine and might create an unusual tilt or sharp angle that may get progressively worse as the individual ages. This is called *hemivertebra*, and some exceedingly rare instances have been recorded in the fossil record.

So far, hemivertebrae have been identified in some ancient amphibians from the Permian and Triassic Periods and in at least one early reptile from the Permian. A single occurrence was documented in a Late Jurassic plesiosaur and a dinosaur, along with a probable record in an embryo or newborn Cretaceous dinosaur and a possible case in a salamander. Only one example has been found in a fossil mammal, that of an Oligocene nimravid (cast your mind back to the false saber-toothed cats in chapter 7). In 2017, an additional case was confirmed in another prehistoric reptile, this time representing the oldest record of hemivertebra in an aquatic animal.

The animal in question is *Stereosternum tumidum*, a type of aquatic reptile, or parareptile, called a mesosaur. The mesosaurs, unlike the similar sounding mosasaurs, were small aquatic reptiles that lived during the Early Permian; they are thought to represent the earliest aquatic reptiles, appearing long before anything like mosasaurs, ichthyosaurs, or plesiosaurs. *Stereosternum* fossils are known from South America and Northern Africa, and this one comes from the State of Paraná in southern Brazil and is around 290 million years old.

The fossil is represented by a virtually complete, approximately 80-centimeter-long skeleton preserved in almost full articulation and exposed in ventral view showing the underside of the skeleton. This view provided the perfect opportunity to inspect the anatomy of the spine, which revealed an unusual malformation in one of the dorsal vertebrae.

Wedged between dorsal vertebra numbers 17 and 19 is a right-sided, trapezoid-like bone identified as dorsal vertebra number 18, which had failed to develop properly. CT scans revealed that both the centrum and neural arch of the pathological vertebra were missing. The distinction in size, shape, and position undoubtedly identifies this as resulting from hemivertebra and is thus a feature that was set from birth. It is also closely in contact with the preceding vertebra (17), although the boundaries of both are distinct. The rib associated with the hemivertebra is noticeably thinner than the other ribs and is angled slightly out of position. The missing piece of the hemivertebra on the other side of the spine also lacks an associated rib. Because of the hemivertebra, both vertebrae 17 and 19 are asymmetrical and appear to be tilted and flattened.

The wedge-like nature of the hemivertebra resulted in the subsequent vertebra, number 19, deviating from its natural position in the spine, so much so that it is noticeably offset with a distinct angular change. This led to the spine curving to the right, then to the left, forming an S-like shape. This is characteristic of S-shaped scoliosis, caused by the spine adjusting to compensate for the hemivertebra's shape, thereby restoring the correct orientation of the vertebral column. As a result, and by this very definition, it means that the hemivertebra can be classified as what is known as incarcerated.

It has been shown that hemivertebra may lead to weakness in the rear limbs and even paralysis in living species. What does this mean for our mesosaur? The mid-column position of the abnormality would certainly have limited the flexibility of the spine, presumably hindering the animal's

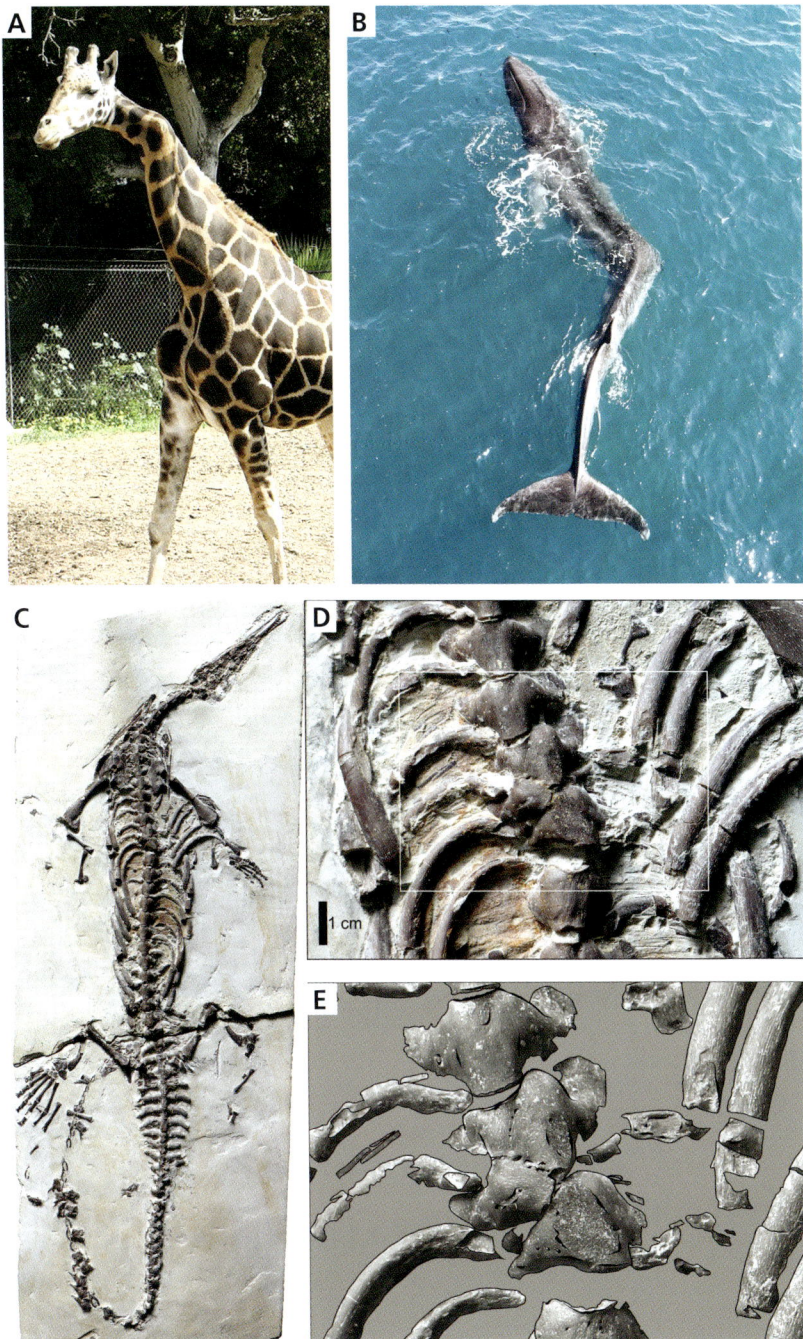

FIGURE 10.8. (A) A giraffe affectionately known as Gemina with a crooked neck and (B) a fin whale with a twisted tail are both possibly the result of scoliosis. (C) The *Stereosternum tumidum* skeleton with an offset, S-like spine typical of S-shaped scoliosis. (D) A close-up of the spine with the wedge-shaped hemivertebra (18th dorsal vertebra). (E) A closer view of the bones outlined in (D).

movement and hunting abilities. The individual, however, represents an adult rather than a juvenile, showing that it survived through adulthood with this condition.

The hemivertebra had seemingly no noticeable impact on this *Stereosternum*, who managed to live a successful life, including the bonus of avoiding being eaten. This might have had something to do with its feeding habits because studies have revealed that mesosaurs mostly dined on small, slow-moving crustaceans and scavenged the occasional carcass, so this presumably alleviated any potential side effects from scoliosis. Further still, mesosaurs are thought to have been quite slow swimmers and relied mostly on their long tails for propulsion rather than lateral undulation of the body.

This fossil is one of the very few confirmed cases of hemivertebra in a complete, articulated skeleton. Having the skeleton provides direct evidence of how the malformed vertebra impacted the rest of the spine and the body. This extremely rare fossil provides the oldest evidence yet of scoliosis, confirming that this condition affecting millions of people worldwide today goes back more than a quarter of a billion years.

FIGURE 10.9. (Opposite). *The Crooked Swimmer*

Two *Stereosternum tumidum* reptiles swim up to investigate a dead coelacanth floating at the surface. One is characteristically symmetrical but her slightly larger companion has a crooked torso and a limp hind limb.

BOB NICHOLLS ART

THE ICE ICE BABY

It is July 10, 2007, and seventeen-year-old me is sitting having breakfast and watching the news on the TV, patiently waiting to capture a glimpse of the latest major paleontological discovery. As I put down my bowl of dinosaur cereal and turn up the volume on the TV, at last, here it is. The breaking news of a sensational mummified mammoth calf, possibly representing the best preserved and most complete woolly mammoth (*Mammuthus primigenius*) ever found.

For any youngsters transfixed with fossils in the 2000s, to get our fossil fix we did not necessarily have the luxury of social media to keep up to date with the latest and greatest fossil finds. Rather, we heard about finds through the radio and TV and in newspapers, with the occasional thing on the internet (if you knew where to look). Of course, we had books, but we had to wait for them to come out, by which time such discoveries were already old news. ("Don't you mean extinct?") As my family were more than aware of my passion for paleontology, they often saved newspaper clippings and, in the case of my beloved Nan, excitedly rang the house phone to tell me that a new fossil find was being shown on the TV. I would drop whatever I was doing and frantically try to find the right channel.

When this woolly mammoth find was announced, it felt like every major media company in the world was covering the story and wanted a piece of the action. There was a huge media storm, which I remember well as I tried to catch as many interviews as possible. If you have not yet guessed, the mammoth calf in question was the lovely Lyuba, found in the remote, icy-cold region of northwest Siberia in Russia's Yamal Peninsula.

While searching for wood along the banks of the frozen Yuribei River in May 2007, Yuri Khudi, a local Nenet reindeer herder, and his sons stumbled upon something special emerging from the permafrost: the frozen body of a female woolly mammoth calf. Exceptionally preserved and about the size of a large dog, Lyuba, whose name means *love* in Russian and who was named in honor of Yuri's wife, was practically complete, with her skin and traces of fur preserved. She looked almost like a modern-day elephant calf, preserved so well that you could imagine she had died just a few weeks ago.

Upon making the discovery, rather than disturbing the carcass, largely in part because many Nenets believe that touching a mammoth is a bad omen or curse, Khudi returned home, informed the local authorities of the discovery, and reported it to the Shemanovskiy Museum, which subsequently led a trip to collect the mammoth. Khudi realized the importance of the find and wanted to ensure it could be taken into the museum's care.

In the following days, a series of unfortunate events occurred, starting with the disappearance of the Ice Age baby. As it turned out, another individual had found the carcass and taken it to a nearby village. Somewhat mysteriously, with the help of the police and even the army, it was later found propped up against somebody's house where, unfortunately, dogs had gnawed off part of Lyuba's right ear lobe and most of her tail. Police confiscated the calf, and she was later flown by helicopter to the Shemanovskiy Museum before heading to a university in Tokyo for further analysis.

We do not need to cover Lyuba's entire backstory, but the point is that she sparked so much curiosity because of her extraordinary preservation and completeness. She was truly an exceptional find. Over the following years, various TV documentaries incorporated Lyuba's discovery, including a *National Geographic* special dedicated to her history and research. As part of the investigatory work, it was later found that another reindeer herder, Nikola Serotetto, had come across Lyuba's carcass a few months earlier, in September 2006, but he had not reported the find.

One of the most exciting revelations about Lyuba was uncovered following a series of CT scans and careful autopsies; this included a full-body scan in 2010, undertaken with the Ford Motor Company in Michigan, which used the same scanner their engineers use to analyze car parts. These scans provided a detailed look at the skeleton and revealed internal organs, including her liver, heart, intestines, and more, providing new and exciting information about mammoth biology. A large hump was also identified on the back of her neck that was made of brown fat, a specialized tissue involved in nonshivering heat production. The fatty hump might have functioned as some sort of thermoregulation, the first time such a feature had been reported in a mammoth.

Rather incredibly, Lyuba's last meal, comprising remnants of her mother's milk, still remained inside her gut. Further analysis of her intestinal contents showed that she had been consuming fecal matter, evidence pointing to a behavior observed in modern elephants where young calves engage in coprophagy and ingest their mother's feces to introduce essential bacteria that aid digestion. Everything about Lyuba showed that she was a normal, healthy individual of her species, consuming small meals of mostly milk, with the occasional act of coprophagy. But . . .

Now comes the tragic part of the story where we discuss the inevitable demise of our beloved baby mammoth. For one thing, Lyuba lived during the Pleistocene, walking the Arctic tundra about 41,800 years ago, leaving her final footsteps in the spring. Rather sadly, based on rigorous analyses of her teeth, Lyuba was just thirty to thirty-five days old when she died. It gets worse. The CT scans that had revealed so much about her life history also revealed how her death played out. Lyuba's mouth and part of her trunk were full of mud, as were her trachea and bronchi, the airways that connect the trachea to the lungs, painting a grim picture of her final hours.

Based on this evidence, it is thought that a fatal accident led to her getting trapped inside thick, sticky mud, perhaps resulting from her mother's herd crossing a muddy lake bed. Panicked and calling out, she attempted to pull herself free from the mud but struggled to escape. Frantically gasping for air, she struggled to breathe as thick mud obstructed her airways, tragically suffocating her.

Where was the mother mammoth, you might ask? Why did she not come to the rescue like we might see in modern elephants that often, but not always, rescue their calves from sticky situations? Maybe Lyuba's mother really was there, struggling to pull her free, or perhaps she did not even see her slip away into the mud. Given the time of year, during the spring, it is also feasible that mother and child passed across thin ice over a lake, where both fell into the frigid waters beneath. There are many unknowns that we will never be able to determine.

Turning this heartbreaking scenario somewhat on its head and looking for a positive, the very same frosty mud that entombed her also contained the perfect recipe for her preservation. Thus, rather than disappearing into oblivion, Lyuba traveled into the future. As she was buried quickly in the fine sediment, this sealed off oxygen. Soon after death, specialized lactic acid–producing bacteria entered her body and preservation began. She then remained frozen in the permafrost for many thousands of years, until one day she became dislodged and wound up in the path of Yuri Khudi.

Lyuba is not the only woolly mammoth calf known. Before her discovery, five other mammoth calves had also been found in Siberia; most famously, the first one, a male called Dima, was found by gold miners in July 1977. Like Lyuba, Dima also had a sad, short life, although he lived for about eight months. In his case, evidence shows that Dima also died from asphyxiation due to the clogging of sediment found in his trachea and bronchi, suggesting that he also died due to the inhalation of mud while struggling to escape from some type of entrapment. Theories as to the cause range from him having fallen through a crevasse in the ice to similarly being stuck in a muddy lake bed. His emaciated body and studies of his empty stomach revealed that he was suffering from starvation. Further mammoth mummies have been found since Lyuba's discovery.

FIGURE 10.10. The author with a perfect replica of Lyuba, the exceptional mammoth mummy calf, at the La Brea Tar Pits in California.

(Courtesy of Natalie Lomax)

As one of the greatest ever finds of an Ice Age animal, Lyuba's story, from her deadly end to her unlikely discovery, captured so many people's hearts. The fateful circumstances that led to her ill-timed and premature death are inherently sad. But her untimely death provided new life, presenting scientists with an exceptional window into the icy world of the best-preserved woolly mammoth mummy in existence.

For me, one of the most uplifting parts of Lyuba's story was Yuri's involvement, not least in finding Lyuba, but how the research team kept him actively involved along the way. They even invited Yuri to see and take part in the autopsy, to help him understand the importance of the discovery he had made. In the primary paper detailing Lyuba's story, published in 2012, it highlighted the meaning of this find for the Nenets. In their worldview, mammoth are seen as creatures of an underworld from which they sometimes might escape, only to die as they enter onto the plane of existence. Lyuba's fortuitous discovery brought two different worlds together, all for the better of science.

Lyuba was toured across the globe as part of a special traveling exhibition, popping up in China, Australia, and the United States, among other countries, to educate the public about this wondrous discovery. In 2014, Lyuba was briefly in London, where I had planned to finally make her acquaintance. Alas, my timing was off, and I never came face-to-face with this beautiful mammoth. Still, I have been able to see some cool replicas. One day, for sure, I hope to see the real deal. Much like how she captured my imagination as a teenager, long may her story continue to inspire new generations.

As a final note, something bizarre happened while sitting at my PC writing about Lyuba. My wallpaper is set to display a revolving slideshow of thousands of photos, mostly of fossils, as you might expect. But, genuinely, as I saved the first draft of this text, the background changed to a photo of me standing with a replica of Lyuba at the La Brea Tar Pits in Los Angeles. An eerie coincidence for sure. Clearly, Lyuba works in mysterious ways.

FIGURE 10.11. (Opposite). *Lyuba's Last Sunlight*

Although the weather is mild, there are still dangers on the chilly steppes. A herd of woolly mammoth (*Mammuthus primigenius*) passed through a boggy lake bed, and their feet churned it up into a deadly trap for little Lyuba. Trapped and left by the oblivious herd, even her agitated mother, Lyuba will go down with the sun.

11 | Stranger Things

Turtles have specialized salt glands in the corners of their eyes to remove salt from their bodies. In the Amazon, when the yellow-spotted river turtle begins to flow tears of salty water, butterflies may be seen drinking their tears as a source of precious salts. This symbiotic, tear-drinking behavior even has its own name—lachryphagy.

BONEHENGE

If you have any interest in prehistory or the slightest bit of knowledge about prehistoric monuments in the United Kingdom, then you have heard of Stonehenge. As one of the most famous landmarks in the UK, Stonehenge is known worldwide for its iconic structure and is considered one of Europe's most important prehistoric landscapes.

Fossil fans, do not fear. We are not about to disappear into the realm of archeology, so do not skip this section. Rather, this is our perfect segue into an unusual site discovered in Russia, nicknamed "Bonehenge."

As a UK native, I grew up learning about Stonehenge in school, though my first visit to the site was in my early twenties. Accompanied by two friends from the United States, we were heading south to study ichthyosaurs from the Jurassic Coast. Although it was a slight detour, they were keen to tick Stonehenge off their UK bucket list—it was fun for me, too. Though it was a whistlestop trip, little did I know that I would be back in just a few years to stand between the stones at sunrise and watch as the sun slowly crept through the gaps in the stone circle. The distinct lack of sound, combined with the steady sunrise and the shadows cast by the stones, created a rather serene atmosphere that quickly made me feel pensive. It was a lovely moment. That quickly changed, however, as we were shortly attacked and chased by British *Velociraptor* relatives called *Nuthetes*. In all seriousness, we were filming at Stonehenge as part of my TV series *Dinosaur Britain*, blending some of Britain's rich dinosaur history with its cultural monuments.

Much of our fascination with Stonehenge revolves around its mystery. The same can be said for Bonehenge. Located adjacent to the Don River, close to the city of Voronezh, a mammoth discovery was made in 2014 at a site referred to as Kostenki 11, a site already well known for the association of mammoth-bone features. Following three excavation seasons, a team unearthed a huge circular structure measuring 40 feet in diameter and built entirely from the bones and teeth of at least sixty-four individual woolly mammoth. This enormous ring of bone is an unusual example of ice age architecture constructed by early humans around twenty-five thousand years ago.

Many mammoth-bone buildings, including much smaller circular structures, have been discovered across Eastern Europe and are well known to researchers. Most of these sites are thought to be the remains of ancient dwellings once used as a form of shelter during harsh glacial winters or perhaps as year-round homes. Some of the most famous examples include huts made from mammoth remains. Similar sites were found at Kostenki in the 1950s and 1960s but were nowhere near the size or extent of the recent discovery.

A huge amount of effort certainly went into building such a large structure, yet what exactly was this giant mammoth ring used for? In a comparable manner to Stonehenge, where many theories have been put forward to interpret the site, with most seeming to agree that it was built as a burial site and some form of prehistoric temple, there are various ideas mixed with uncertainty surrounding Bonehenge.

Primarily, the size of the site is unlike any similar mammoth-bone structure found so far. That alone makes it unusual and implies that there was surely a reason for it being so large. Investing significant time and energy into making something this big would suggest that it was made to last. The first thought might be as a type of huge shelter, though this is unlikely given the challenges of covering such a large structure; there is no specific entrance to the circle, either. The site also lacks little evidence of any other prey animal remains, which are usually found in association with smaller mammoth-bone structures. This would probably imply that people were

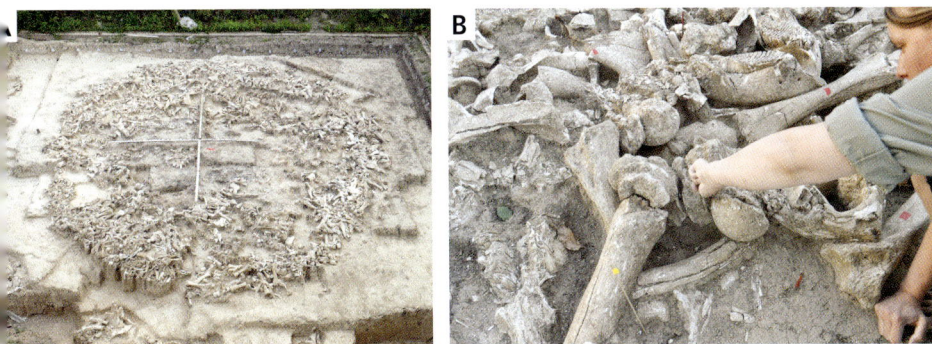

FIGURE 11.1. Kostenki 11, nicknamed Bonehenge. (A) Photograph of the large circular mammoth-bone structure. (B) A close-up of some of the mammoth bones, including jaws with teeth.

([A] Courtesy of Sasha Dudin, 2017; [B] courtesy of Alex Pryor, 2015)

not living there as a dwelling. However, burnt bones and charcoal were also found, showing that humans were at least making fires.

A research team led by the archeologist Alexander Pryor of the University of Exeter published a study of the site in 2020 and considered some possibilities. Based on previous interpretations of similar, albeit smaller sites, they mentioned the possibility of the structure representing some sort of monument or ceremonial feature; however, they suggest that the most likely function was as a food storage facility, sort of like a giant larder of mammoth meat. This interpretation might render most of the structure as representing an unusual midden, a huge heap of animal waste thrown away by the hunter-gatherers, though the creation of such a precise circle means this is probably unlikely.

Some support for the storage interpretation is the discovery of hundreds of flakes and fragments (called debitage) left behind from the creation of tools, including evidence of knapping, indicating that these tools were used to process the meat. The structure was also associated with three pits found just outside the main bone ring. These 1–2-meter-diameter pits contained large mammoth bones and were presumably used to store the meat.

A question remains as to where all these mammoth came from and how they were carried to this exact spot, a tall task in any circumstance but made even tougher in the icy cold and hostile conditions. Presumably, some must have been hunted for their meat, especially if this site was indeed for food processing and storage, although the presence of so many mammoth suggests that the hunter-gatherers must surely have scavenged, too. Perhaps they recovered carcasses of animals that died from natural causes, maybe unfortunate members of herds that perished while crossing a nearby river. Whatever the exact reason for the construction of this strange mammoth monument, it makes for one of the most bizarre circular structures created by early humans who utilized the largest mammals of their time, the mighty mammoth.

FIGURE 11.2 (Opposite). *Bonehenge Boom Time*

It is the busiest time of the year at Bonehenge. A tribe of people (*Homo sapiens*) have already taken advantage of a passing herd and are butchering a woolly mammoth (*Mammuthus primigenius*). While some bury slabs of mammoth meat, others make weapons and collect firewood.

LIFE'S A DRAG

When my study describing an extraordinary 9.7-meter-long horseshoe crab death track was published in 2012, I felt a great sense of pride, considering that it was such an exceptional Jurassic fossil with a remarkable story locked in time. It was genuinely *the* specimen that changed the way I thought about fossils, specifically in how remarkable evidence of behaviors might be preserved. It was also one of my first academic papers, so admittedly, it was pretty neat.

As any active publishing scientist will tell you, and if you are an academic reading this then you will understand, once you publish your research, you go through stages of worrying whether people will agree with your findings and, of course, if anybody will even read and reference your work (impostor syndrome overload). Over the next couple of years, the study was cited by various researchers, and in 2015, somewhat prompted by another citation of the paper, I decided to browse the internet to see if any similar specimens had come to light.

While searching for things like "horseshoe crab death track" or "fossilized trackway Solnhofen," I happened by chance to come across a photo of a long and unusual track that seemed to have an ammonite at the end. I pondered whether this could be an ammonite death track—or mortichnion—and was inspired to examine it. In October 2015, I reached out to the Cosmo-Caixa Museum in Barcelona, Spain, where the specimen was displayed, and inquired about studying the fossil. Unfortunately, my initial email was lost in the void, and it would not be until the following October before I would eventually get to see the fossil in person, following an exchange of exciting conversations with the curator, Alex Jiménez.

The specimen was found in the late 1990s, cleaned and prepared in 1998, and acquired by the museum in 2002. It was collected from the famed Late Jurassic, 150-million-year-old Solnhofen limestones of southern Germany—like the horseshoe crab trackway—and probably came from a quarry in the Langenaltheim Haardt district, near the village of Solnhofen.

When I arrived at the museum on Monday, October 17, Alex had prepared the display for me, ensuring that the fossil was accessible and that the extensive glass casing had been removed. The first thing to do was to measure the apparent "track," which, from the preserved first section to the base of the ammonite, measured a staggering 8.5 m! The track is mainly straight, with a few minor changes in lateral direction. It is comprised of continuous parallel ridges and furrows, initially starting with just two prominent ridges and a single furrow, with a width of 5.7 millimeters, but more ridges and furrows gradually become visible along the length of the track. Beyond about the 7.5-meter mark, five to six prominent ridges are present and at 3 centimeters from the ammonite, the number of ridges increases to eleven.

What does this show? Could it be that these ridges represent a trace made by the arms or tentacles of the ammonite as it swam along the bottom, or maybe they show the eternal struggle of a dying ammonite pulling itself along? In reality, neither of these scenarios is correct. The animal was not alive when this "track" was made. As such, this fossil does not represent or capture evidence of behavior, and this is not technically a track because a living animal did not create it. Instead, it is a drag mark, otherwise referred to as a "sole mark or tool mark," produced in this case by the animal's shell after its death.

It probably seems odd to think about this, but dead animals may leave behind some sort of markings or surface structures, all dependent on the environment in which they died and what their body was subjected to afterward. As it turned out, some simple drag marks had been reported from the Solnhofen limestones previously, including those made by jellyfish, driftwood, and indeed ammonites. Interestingly, even ammonite roll marks have been reported, which conjures an odd image of empty ammonite shells rolling along in the current. Given the unusual nature of these ammonite markings, the first accounts published in the very early 1900s were initially mistaken as trace fossils, such as tracks or scratch marks, left behind by vertebrates such as turtles or fish.

A few drag marks from the Solnhofen limestones have been documented with associated ammonites, though each mark is less than 1 meter long. Speaking of which, the creator of this epic 8.5-meter mark is an ammonite in the family known as perisphinctids, and is called *Subplanites rueppellianus*, a common species found in the Solnhofen limestones. The specimen represents a subadult male and measures just 11.4 × 10 centimeters, yet it created a staggering 8.5-meter drag mark. Okay, so what gives? How could

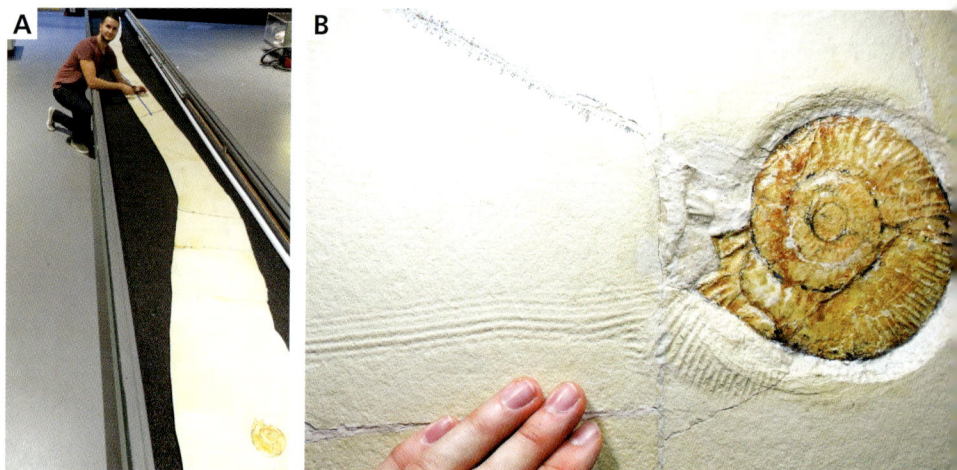

FIGURE 11.3. (A) The author studying the entire 8.5-m-long ammonite drag mark in 2016. (B) A close-up of part of the drag mark and the ammonite *Subplanites rueppellianus*.

(Photographs by the author)

a dead ammonite leave its everlasting mark, and were the ridges made by the dead animal's arms?

The drag mark was made by a dead, floating ammonite whose shell ribs penetrated the surface of the ancient seafloor as it gradually swayed along in a very calm but constant current. Shells of dead ammonites may also be moved along the substrate by waves and winds. The ammonite probably only recently died and remained buoyant with the aid of decaying gases contained inside the ammonite's chambers, which were filled with gas during life and controlled their buoyancy and movement, a bit like a

FIGURE 11.4. (Opposite). *The Ammonite's Last Move*

A metriorhynchid, a marine relative of crocodiles, came from nowhere and skillfully bit off the lifeless soft parts of the ammonite *Subplanites rueppellianus*. Still filled with decaying gases, the empty ammonite continues to gently drift on the current before arriving at its final resting place.

submarine. The individual ridges and furrows were created as the ribs were pulled along, with more ridges present when more of the ammonite was in contact with the substrate.

Eventually, the last bit of gas from the ammonite's ghost escaped from the shell, only for it to fall over onto its side and remain forever in that position. Right at the point at which the ammonite is preserved, eighteen ridges (formed by the ribs) can be counted, and their orientation shows that the ammonite touched the substrate and rotated slightly before coming to rest.

Though evidence of trackways and tracemakers preserved together in the fossil record is rare, the co-occurrence of a drag mark preserved with the dead animal that produced it is rather exceptional—so much so that the fossil represents the hitherto longest fossil drag mark created by a dead animal, complete with its maker preserved at the end. The explanation is not quite zombie vibes, but it captures an unusual moment in the afterlife of this ancient mollusk.

THE ZOMBIE DEATH GRIP

No sooner had plants spread far and wide across the planet did insects begin munching, moving, and making their homes on them. Rather than allowing their leaves to be lost to all these behavioral interactions, some plants fought back, developing defensive features, from spikes and thorns to toxic leaves. Although it is impossible to say when the first prehistoric plants began to be exploited in such a way, enormous fossil forests paved the way for plants and herbivorous invertebrates to thrive and dominate terrestrial ecosystems for over four hundred million years.

The plant fossil record is filled with astounding, beautiful fossils that reveal many signs of invertebrate interactions. Most of these are made by insects, leaving behind evidence of feeding, leaf mining, and galls, but these stories are better left for another time and place. Here, we will focus on something a little more, should I say, macabre?

To set the scene a little, we shoot back in time to an extensive lake in the heart of a lush tropical forest in the Eocene, about forty-eight million years ago, and to one of the most famous fossil sites of the time, Messel. Famed for its beautifully preserved fossils, including pregnant fox-sized horses and mating turtles, the Messel Pit near Frankfurt, Germany, has revealed an incredible diversity of flora and fauna with insects among the most frequently found fossils.

Our object of interest is a single leaf of a dicotyledonous plant, or dicot, called *Byttnertiopsis daphnogenes*. Yes, I know that is a bit of a mouthful, and I struggled to pronounce it, too. Dicotyledonous plants sound like they might be rare or unusual, but they are hugely varied with over two hundred thousand species, including many familiar flowering types. In fact, the sweet potato (*Ipomoea batatas*) is a type of dicotyledonous plant, of which we eat its nutritious tuberous roots. This Messel *Byttnertiopsis* leaf preserved a unique fossil association: leaf scars produced by an ant.

An ant-bitten leaf. So what, right? We have ants eating leaves outside our windows right now. The significant thing about this specimen is what the shape of the leaf scars tells us. The leaf is nearly complete, measuring just under 10 centimeters, and has twenty-nine distinct dumbbell-shaped scars centered on eleven secondary veins. This scar pattern is stereotypical of "death-grip scars" on damaged leaves made by ants that were hijacked by a fungal infection. There are many fungi today that manipulate insects to

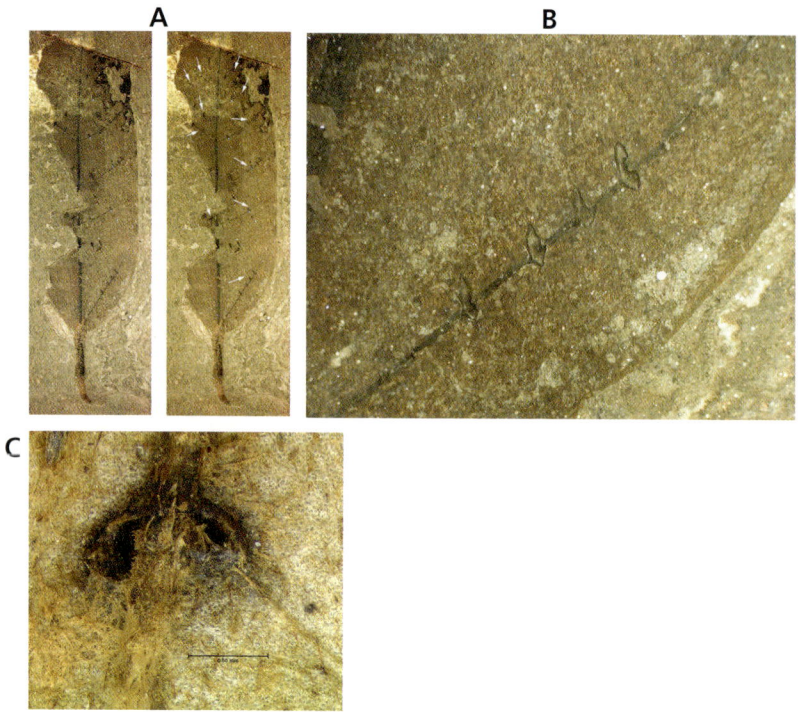

FIGURE 11.5. (A) The nearly complete fossil leaf *Byttnertiopsis daphnogenes*, with twenty-nine ant death-grip scars centered on eleven secondary veins (arrows). (B) A close-up of four scars on the lowermost vein; note the dumbbell-like-shaped holes. (C) Modern ant death-grip scar from a primary-secondary vein showing the dumbbell-shaped hole.

([A] Courtesy of Torsten Wappler; [B–C] courtesy of Conrad Labandeira and David Hughes)

bite leaves, become zombified, and do the fungus's bidding. As death-grip scars are so distinctive, they are plant-insect interactions that would seem plausible to preserve in fossil leaves. This is the first.

Ants whose brains were taken over by fungi. Correct. It might sound like a postapocalyptic *The Last of Us* type of science fiction, but many parasites

FIGURE 11.6. (Opposite). *The Killer Reveals Itself*

Although it has been dead for several days, the ant's jaws remain clamped to the leaf (*Byttnertiopsis daphnogenes*). Although it was a nightmare end for the ant, the parasitic fungus that sprouted from its head is a beautiful treat for everyone else's eyes.

BobNichollsArt

today have evolved abilities to manipulate host behavior, and this dramatic example is observed in the modern world when fungi force insects to die attached to leaves. The most striking evidence, and the best-documented interaction of this type of behavior, is found among worker ants of the species *Colobopsis leonardi* in Thai tropical forests that become infected by the fungus *Ophiocordyceps unilateralis*. Funnily enough, if you are familiar with *The Last of Us*, it is indeed *O. unilateralis* that led to the fungus-infected humans!

In a gruesome twist of fate for the ant, it becomes infected with parasitic fungus spores that infiltrate its body and grow beside its brain, altering its behavior and forcing it to climb down the canopy to a specific height of about 25 centimeters above the forest floor, an ideal spot for spore release. At this point, the ant is no longer an ant but has been taken over by the fungus that is controlling its behavior. The zombie ant *lives*. Only then, under the fungus's spell, using its tough mandible, it firmly locks itself via a death-grip bite onto a major vein of a leaf. Following the ant's death due to the fungal parasite, the fungus drains the last dregs of nutrients from the ant before a long stalk gradually bursts up through the host's lifeless head and begins to disperse its spores, ready to infect more ants and restart the deadly cycle.

One study in southern Thailand found that the fungus was very specific about which ant host it wanted, with 97 percent of hosts being a single carpenter ant species. The telltale marks left behind by the ants are a pair of widened puncture marks made by the mandibles of the ants, just like the dumbbell-like shapes in the fossil leaves. Although we do not know which type of ancient ants were brainwashed by the fungi at Messel, carpenter ants of the genus *Camponotus*, which the *O. unilateralis* fungus targets today, date back at least fifty million years. Though none have been found at Messel, *Camponotus* fossils have been found inside similarly aged amber deposits in Germany, Ukraine, and the United States, suggesting that perhaps the same genus may also have been present in Messel.

The rather unassuming leaf from forty-eight million years ago records evidence of ant-fungal parasitism, showing that this highly specialized interaction has its roots much deeper in time. At least twenty-nine ants were brainwashed into playing host and succumbed to a fungal parasite. This is an extraordinarily rare and first-known example of behavioral manipulation in the fossil record.

ICE AGE REGENERATION

You might have heard that people are trying to "bring back the woolly mammoth" and other Ice Age creatures. It might sound like something out of science fiction, but over the last decade or so, there has been a bit of a media blitz about de-extinction, the process of bringing extinct species back to life. The core focus or poster child of this process has been the woolly mammoth because several wonderfully preserved carcasses have been found frozen in the permafrost, such as Lyuba, and the "mammoth resurrection" story periodically does the rounds on social media.

I am sorry to disappoint those who might wish to see a living woolly mammoth or maybe try some mammoth jerky because this *is* science fiction. If somebody were to recreate a so-called mammoth, it would not actually be a real mammoth but just some oddball, perhaps slightly hairier elephant hybrid. As I have said before, and will happily say again, I would rather that the time, money, and research went into saving animals that are at risk of extinction today.

It is not just mammoth that have been found frozen in the permafrost; other ice age beasties include woolly rhinos, bison, cave lion cubs, saber-tooth cat kittens, and more. None of these can truly be brought back to life, though one little ice age mammal inadvertently helped future scientists achieve something close to it.

That little mammal was a ground squirrel, and its helping hand was a burrow it had created some 31,800 years ago. Discovered in northeastern Siberia on the bank of the Kolyma River, this burrow was one of at least seventy burrows (many containing nests and caches, together called middens) found buried at depths of 20–40 meters that remained undisturbed inside frozen permafrost sediments. The burrows were discovered in layers formed under tundra-steppe conditions that contained the bones of other contemporary mammals, including mammoth, woolly rhinos, bison, horses, and deer, as well as plant remains.

The type of burrow-making squirrel is thought to be the same species of living Arctic ground squirrel called *Urocitellus parryii*, which is found today in icy conditions in Alaska, parts of Canada, and, indeed, Siberia. However, an exact identification is difficult because no carcass was found in the burrow, and other ice age species are known, including mummified individuals. In any case, the ancient squirrels that created these burrows were either early examples of today's Arctic ground squirrel or something very close to it.

The fossilized burrows had never been defrosted following their initial burial and proved to be the perfect time capsule. As part of a 2012 study, more than thirty of the seventy burrows were investigated, and many contained plant remains. Much like modern Arctic ground squirrels, but also marmots, groundhogs, and chipmunks, these ice age squirrels had collected a cache of seeds and fruits that they had carefully stashed inside their burrows.

The storage chambers inside the burrows contained a huge supply of food, with a staggering six to eight hundred thousand frozen seeds and fruits! When the burrows were constructed, the chambers containing the seeds and fruits were built up against the frozen boundary of the permafrost, which enabled their perfect preservation. Considering that so many seeds and fruits were perfectly preserved in a seemingly cryogenic state, could they have remained fertile?

Previous attempts to regenerate ice age seeds into fully-fledged plants had been undertaken without success. Somewhat miraculously, this team of researchers were not only able to successfully grow some of the fossil fruits into mature plants, but those plants also produced flowers, fruits, and viable seeds.

The successful regeneration focused on the fruit tissues of a small perennial herbaceous plant, which the team identified as *Silene stenophylla*, also known as the narrow-leafed campion. Another team later determined it to be the species *Silene linnaeana*. There are more than eight hundred known species of

FIGURE 11.7. (Opposite). (A) A thirty-thousand-year-old mummified Arctic ground squirrel, *Urocitellus parryii*, collected from northeastern Siberia. (B) Representative Arctic ground squirrel frozen midden, including (C–D) some close-ups of huge caches of seeds and fruits from the twenty-five to thirty-thousand-year-old mammoth steppe of Yukon, Canada. (E–F) Fruiting plants of *Silene* regenerated from the tissue of fossil fruits, and a regenerated plant from the tissue of fossil fruit with both female and bisexual flowers.

([A] Image from M. Faerman, et al., "DNA Analysis of a 30,000-Year-Old *Urocitellus glacialis* from Northeastern Siberia Reveals Phylogenetic Relationships Between Ancient and Present-Day Arctic Ground Squirrels," *Scientific Reports* 7 (2017): 42639; [B–C] images courtesy of Grant Zazula and the Government of Yukon; [D] courtesy of Scott Cocker; [E–F] image from S. Yashina et al., "Regeneration of Whole Fertile Plants from 30,000-y-old Fruit Tissue Buried in Siberian Permafrost," *PNAS* 109 (2012): 4008–13)

Silene, many of which bear numerous and often very subtle similarities, so it is easy to understand the difficulties with providing an identification. Regardless, both species still grow today in the Arctic tundra of Siberia.

The research was focused on samples that were preserved as a conglomerate of *Silene* seeds and fruits collected from inside a single fossil burrow found at a depth of 38 meters. Separately, the team used seeds from extant species from the same region to compare the growth with the ancient examples. Curiously, as the ancient and extant plants grew, they turned out to be slightly different, with the ancient group producing one and a half up to three times more buds and the extant plant producing roots more rapidly. The team grew a total of thirty-six plants from the fossil fruit, each of which was identical. Upon flowering, subtle differences were apparent between the ancient and modern plants, including in the shape and length of the petals and the sex of the flowers.

This ancient resurrection is by far the oldest plant material brought back to life. These ice age squirrels unknowingly provided the closest thing to seeing a real "Ice Age Park," albeit with plants. Who would have thought that such a scenario would have been possible due only to the careful collecting and hoarding behaviors of these little squirrels, who squirreled away their food cache only for it to be regenerated some thirty-two thousand years later? A circle of life moment would have been if these scientists had then fed the reproduced seeds to the squirrel descendants living in the same area. Maybe they have, or maybe this is something that a reader might one day carry out.

If you are familiar with the *Ice Age* film franchise, then it is quite pertinent to mention the character Scrat, that saber-toothed scraggly squirrel who is always trying to stash his acorn somewhere safe. Though fictionalized, Scrat's behavior echoes the real-life behaviors of these ancient squirrels that successfully buried their stash of seeds and fruits, though rather unluckily, and strangely in keeping with Scrat's tradition, they also lost their precious stash. Perhaps this is an unlikely and odd example of life imitating art.

FIGURE 11.8. (Opposite). *The Proud Collector*

A busy Arctic ground squirrel (*Urocitellus parryii*) pauses outside her seed-laden burrow, below swaths of grasses and ancient campions. In the middle ground are a passing herd of woolly rhinoceros (*Coelodonta antiquitatis*).

TWO HEADS BETTER THAN ONE?

In the realms of fantasy, folklore, and mythology, there is usually at least one animal that has two or more heads. From the mighty serpentlike Hydra to Cerberus, the three-headed hound of Hades, King Ghidorah in *Godzilla*, and Exeggutor in *Pokémon*, these multiheaded creatures trigger our inquisitive minds. Though, as the saying goes, "two heads are better than one," alluding to the fact that problems might be better solved by two or more people working together, if any TV show or monster movie has shown us anything, it's that the heads do not always agree or get along.

The unusual phenomenon of having two heads is known as axial bifurcation, and two-headed animals may be referred to as bicephalous. This developmental abnormality or malformation occurs during embryo development. Two heads may result from an embryo beginning to split into two separate individuals but essentially failing to complete the separation. Animals that are born with two heads have a low survival rate and many die at birth, though some may go on to live a relatively normal life. The most famous bicephalous types of living animals include snakes and turtles, with hundreds of documented cases. Many individuals have been shown to live for years in captivity. Rarer still are those of mammals. Given that this unusual happening occurs in living animals, why not in prehistoric species?

Enter one of the coolest fossils ever, a two-headed Cretaceous reptile. More specifically, a choristodere, which you might recall having met in chapter 1. Remember our choristo friend with eighteen embryos? This two-headed individual belongs to the same type, *Hyphalosaurus*, but to a different species, *Hyphalosaurus lingyuanensis*. Like the others, it was collected from roughly 120–125-million-year-old Early Cretaceous rocks of the famed Jehol Biota fossil beds of Liaoning Province, China.

FIGURE 11.9. (Opposite). The only known bicephalous fossil is this example of *Hyphalosaurus lingyuanensis*, a choristodere with two heads and two necks.

(Courtesy of Eric Buffetaut)

The Jehol fossil beds have revealed a plethora of remarkable fossils, from fossilized soft tissue structures to fully feathered theropods, but this two-headed little reptile takes the title of most bizarre. It certainly takes another title, too, being the only known fossil with two heads and two necks. Given that axial bifurcation occurs in the animal kingdom today, albeit still inherently rare, it could be expected that more examples could or should be known from the millions of fossil discoveries, yet this is the sole representative found.

The thousands of known *Hyphalosaurus* fossils provide paleontologists with a wealth of material for study and comparison. When having lots of specimens to examine, subtle features may sometimes be detected, such as an individual with a slightly longer neck or tail, or an additional bone in an appendage, but having two distinct necks and heads is substantially more than a subtle difference. Instead, having thousands of specimens available to study, compared to say just five individuals, means that there is a higher chance that some sort of abnormality might be found, as is the case with our tiny two-headed marvel.

Some skeptics might question whether the fossil is a fake, considering that some famous fossil fakes have come out of China. The specimen has been studied and scrutinized by several paleontologists and was formally described in a prestigious scientific journal in 2006 (I remember reading about this find in a newspaper and still have the clipping). There is zero evidence of any tampering with the fossil, and the two necks are clearly attached to the body at the level of the pectoral girdle. Measuring from the tip of either snout to the end of the long tail, this abnormal (or *teratological*) specimen measures just 7 centimeters long. The two heads and two necks are identical in shape and size and are positioned close to each other in a similar position.

In the original study, the researchers referred to this individual as either a late-term embryo or a neonate. As we know from the pregnant mother,

---

FIGURE 11.10. (Opposite). *The Unfortunate Duo*

Beside a freshwater pond, a freshly emerged, rare two-headed Cretaceous reptile, *Hyphalosaurus lingyuanensis*, enjoys the warmth of the sun. Unfortunately, its first day on Earth will also be its last.

BOB NICHOLLS ART

who was 80 centimeters long, her embryos were very close to being delivered and were around the same size as free-living individuals that have also been found. Small-sized neonates are 6–10 centimeters long. Thus, given the size and the fact that it is not associated with any adult, it is possible that, rather than representing an ejected embryo, this two-headed beastie might have been a very young neonate that perhaps survived for a little while before succumbing to an early death.

Studies focused on the area's geology have shown that the many *Hyphalosaurus* fossils were buried in moderate to deep water and preserved by ash fall, hence their often exquisite preservation. The fact that we have so many individuals, but just one two-headed example, further highlights the extreme rarity of this phenomenon.

In many ways, this little discovery not only symbolizes the fragility of the fossil record but emphasizes its unpredictability. It ultimately highlights that it is not impossible to find other two-headed fossils. Now you are probably wondering if this means that a two-headed dinosaur is somewhere out there; in fact, even a two-headed mutant dinosaur was briefly featured in the 2025 movie, *Jurassic World Rebirth*. Again, the reality is that it is not impossible or in the realm of fantasy, especially considering that extremely rare examples of two-headed living theropods have been documented. Who knows what the fossil record might throw our way in the future? Only time will tell.

This exceptionally unusual fossil, much like the others featured in this book, makes almost anything feel possible in the world of paleontology. Curiosity about the unknown and the wonder of what might be found next in the lost worlds of deep time will continue to supercharge our fascination with this awe-inspiring science, keeping our passion for the prehistoric past very much alive in the present.

Acknowledgments

DEAN LOMAX

First, I want to thank you, dear reader, for picking up this book and diving into the extraordinary private lives of prehistoric animals. What a ride. Who knew that there were so many exceptional and varied fossils that record the real-life behaviours of long-dead creatures? If you have enjoyed this book, which if you have gotten this far, then chances are that you have, please let me (and Bob) know! I have spent years developing and writing this book, and I enjoy reading honest comments, messages, and reviews from readers who have enjoyed coming with me on this journey. It means a lot to me but also encourages renewed interest in paleontology. Your review counts for a lot. Above all, I hope this has opened your eyes to some of the most breathtaking, unexpected fossils ever found. Until next time, happy adventures.

By a remarkable twist of fate, I found myself starting to write these acknowledgments on the four-year anniversary of my beloved mum's passing. In a strange way, this brought me solace, reminding me that even on a dark day, I can find light. It's a chance to reflect on the journey that led to this book and to acknowledge the people who have been there with me every step of the way.

At every opportunity, I will always acknowledge the profound sacrifices that my mum made for me and my siblings. Her selflessness and love continue to inspire me. If it weren't for her, I would not be the person I am today. And for that, I appreciate all she was and everything she allowed me to become. Through her care, love, patience, and everlasting encouragement, she gave me all the support in the world to achieve my desire to become a paleontologist. For that, I appreciate just how lucky I was to have such a caring, loving mum.

As Mum would always say, family is the most important thing. I will forever acknowledge the help of those with me today and those whose souls still inspire me to continue to do better. A huge thanks to the following for your love and support: Joyce Lightfoot, Scott and Karen Lomax, Julie, Mark, Olivia, and Fletcher Boyles, Ken Lomax and Jane Miller, and Russ and Lorraine Turner. I especially want to thank Natalie, Eevee, and Lucie Lomax for *living* this book with me, hearing me recite paragraph after paragraph and enduring endless questions. But, above all, thank you for always encouraging and supporting me. I love you all.

To my dear friend and fellow paleontologist Jason Sherburn, I feel like we have been to Middle-earth and back with this book! By this, I give a huge thanks to Jason for providing a detailed review of the first draft. He offered precious feedback that helped to improve the book. I appreciate all your help and support.

One of the most inextricably wondrous things about paleontology is that almost everybody that you meet who loves fossils—at least who has ever crossed my path—is generally always excited to chat about prehistoric life and help in some way, shape, or form. A five-year-old or a ninety-five-year-old can easily share a special connection through their fascination for fossils. This is one of the many reasons why I love this science so dearly. As such, I have a long list of people who I would like to acknowledge for their help and assistance during the creation of this book.

The book is a culmination of years spent studying and writing about prehistoric (and living) animals and keeping up to date with the latest research. It is based on countless years of dedicated research and fieldwork undertaken by fellow paleontologists and fossil hunters of all ages. I am grateful to the following people for kindly taking the time to chat with me about their research and discoveries, offer encouragement, and provide information that has helped to make this book a reality:

Kathryn Abbott, Mary Anning, Malcolm Bedell Jr., Phil Bell, Mike Benton, Riley Black, Steve Brusatte, Markus Bühler, Keaton Burghardt, Dawn and Matthew Butler, JP Cavigelli, Nicolás Chimento, Karen Chin, Melissa Connely, Michael Cormack, Reece Davies, Paul de la Salle, Danielle Dufault, Mackenzie Enchelmaier (at The Australian Age of Dinosaurs Museum), Steve Etches, David Evans, Mike Everhart, Everything Dinosaur (Mike Walley and Sue Judd), Andy Farke, Brian Fernando (aka "Big Fossil"), Owen Fielding, Richard Forrest, George Frandsen (and the Poozeum), Victor Ghirotto, Dan Goldsack, Angie Guyon and the Wyoming Dinosaur Center, Ashley and Lee Hall, Ellie Harrison, Rolf Hauff, Don Henderson, Eva Hoffman, Sally and Neville Hollingworth, Dave Hone, Elaine, Mary, Sandra and Reid Howard, Jim Kirkland, Nigel Larkin, Jessica Lippincott, Bruce Lieberman, Alex Liptak, Jeremy and Patricia Lockwood, Spencer Lucas, Jordan Mallon, Judy Massare, Erin Maxwell, Andrew Milner, Cassius Morrison, Levy and Kelly Morrow, Darren Naish, Emma Nicholls, John Nudds, Jingmai O'Connor, Elsa Panciroli, Susan Passmore, David Penney, Tony Pinto (and the "Why Dinosaurs?" team), Mike Pittman, Andrew Rossi, the Royal Commission for the Exhibition of 1851, Sven Sachs, Gabe Santos, Beth-Ann and Matt Schulz, Levi Shinkle, Slash, Aaron, Shae, Mark, and Helen Smith, Sally-Ann Spence, Stephan Spiekman, Hans-Dieter Sues, Emily Swaby, Jon Tennant, Helmut Tischlinger, Ray Troll and David Strassman (the PaleoNerds podcast), Jack and Susan Turnbull, Sue Turner, Bill Wahl, Jimmy Waldron, Matt White, Darren Withers, the Western Interior Paleontological Society (WIPS), Greg, Isabelle, Belenda, Jean-Luc, and Rosie Willson, Betty and Warren Withers, Luke Weaver, Mark Witton, Grant Zazula, Zhonghe Zhou, and JP Zonneveld.

Thanks also to Kallie Moore ("and you too can do it") and Filippo Bertozzo for kindly reviewing the first draft of this book and providing tremendously supportive comments.

Last but certainly not least, a huge thanks to the incredible Bob Nicholls. Thanks for being an all-around brilliant and marvellous artist, scientist, and friend but especially for bringing the fossils, their stories, and this book to life in such an extraordinary way, helping readers to connect so incredibly well with the stories that are forever written in stone.

BOB NICHOLLS

My thanks go to Victoria, for doing more than her fair share, not just when I'm working weekends and evenings but all the time. Thanks, Darcey and Holly, for being so good and reminding me what is most important; you're the most brilliant members of Team Brilliant. Thanks, Granny and Grampy, Nanny and Pops, for always being there, too.

Thanks, Gareth, for the tech support, and huge thanks (and sorry) to Duncan, for waiting yet another year, this time for me to finish illustrating this book.

Finally, a big thank you to Dean for collating such fantastic fossils. It's been a pleasure as always. What shall we do next?

We would both like to say "thank you" to our fabulous agent, Ariella Feiner, and also to Miranda Martin and everybody at Columbia University Press. Thank you for all the support and encouragement and for making this process so very enjoyable.

Further Reading and References

PROLOGUE: PALEO IN PERSPECTIVE

Introduction: The Journey Doesn't End Here

Attenborough, D. F. *The Trials of Life: A Natural History of Animal Behaviour.* Collins/BBC, 1990.

Benton, M. J. "Studying Function and Behavior in the Fossil Record." *PLOS Biology* 8 (2010): e1000321.

BBC News. 2015. "Weasel Photographed Riding on a Woodpecker's Back." https://www.bbc.co.uk/news/uk-31711446.

Boucot, A. J. *Evolutionary Paleobiology of Behavior and Coevolution.* Elsevier, 1990.

Boucot, A. J., and G. O. Poinar Jr. *Fossil Behavior Compendium.* CRC Press, 2010.

Hsieh, S., and R. E. Plotnick. "The Representation of Animal Behaviour in the Fossil Record." *Animal Behaviour* 169 (2020): 65–80.

Lomax, D. R. *Locked in Time: Animal Behavior Unearthed in 50 Extraordinary Fossils.* Columbia University Press, 2021.

1. STARTING OUT

Introduction

O'Connell, L. A., and D. Crews. "Evolutionary Insights into Sexual Behavior from Whiptail Lizards." *Journal of Experimental Zoology Part A: Ecological and Integrative Physiology* 337 (2022): 88–98.

The Roots of Reproduction

Butterfield, N. J. "*Bangiomorpha pubescens* n. gen., n. sp.: Implications for the Evolution of Sex, Multicellularity, and the Mesoproterozoic/Neoproterozoic Radiation of Eukaryotes." *Paleobiology* 26 (2000): 386–404.

Droser, M. L., J. G. Gehling, L. G. Tarhan, et al. "Piecing Together the Puzzle of the Ediacara Biota: Excavation and Reconstruction at the Ediacara National Heritage Site Nilpena (South Australia)." *Palaeogeography, Palaeoclimatology, Palaeoecology* 513 (2019): 132–45.

Dzaugis, P. W., S. D. Evans, M. L. Droser, J. G. Gehling, and I. V. Hughes. "Stuck in the Mat: *Obamus coronatus*, a New Benthic Organism from the Ediacara Member, Rawnsley Quartzite, South Australia." *Journal of Earth Sciences* 67 (2020): 897–903.

Gehling, J. G., and M. L. Droser. "Synchronous Aggregate Growth in an Abundant New Ediacaran Tubular Organism." *Science* 319 (2008): 1660.

McNamara, K. J., and S. M. Awramik. "Stromatolites: A Key to Understanding the Early Evolution of Life." *Science Progress* 76 (1992): 345–64.

Surprenant, R. L., J. G. Gehling, and M. L. Droser. "Biological and Ecological Insights from the Preservational Variability of *Funisia dorothea*, Ediacara Member, South Australia." *PALAIOS* 35 (2020): 359–76.

Mega Millipede Mating

Brookfield, M. E., E. J. Catlos, and S. E. Suarez. "Vertebrate Lies? Arthropods Were the First Land Animals!" *Geology Today* 38 (2022): 65–68.

Davies, N. S., R. J. Garwood, W. J. McMahon, J. W. Schneider, and A. P. Shillito. "The Largest Arthropod in Earth History: Insights from Newly Discovered Arthropleura Remains (Serpukhovian Stainmore Formation, Northumberland, England)." *Journal of the Geological Society* 179 (2021): jgs2021-115.

Jovanović, Z., S. Pavković, B. Ilić, et al. "Mating Behaviour and Its Relationship with Morphological Features in the Millipede *Pachyiulus hungaricus* (Karsch, 1881) (Myriapoda, Diplopoda, Julida)." *Turkish Journal of Zoology* 41 (2017): 1–28.

Lomax, D. R., P. Robinson, C. J. Cleal, A. Bowden, and N. R. Larkin. "Exceptional Preservation of Upper Carboniferous (Lower Westphalian) Fossils from Edlington, Doncaster, South Yorkshire, UK." *Geological Journal* 51 (2014): 42–50.

Wilson, H. M., and L. I. Anderson. "Morphology and Taxonomy of Paleozoic Millipedes (Diplopoda: Chilognatha: Archipolypoda) from Scotland." *Journal of Paleontology* 78 (2004): 169–84.

Whyte, M. A. "Mating Trackways of a Fossil Giant Millipede." *Scottish Journal of Geology* 54 (2018): 63–68.

Giant *T. rex* Penis

Bell, P. R., C. Hendrickx, M. Pittman, T. G. Kaye, and G. Mayr. "The Exquisitely Preserved Integument of *Psittacosaurus* and the Scaly Skin of Ceratopsian Dinosaurs." *Communications Biology* 5 (2022): 809.

Bell, P. R., C. Hendrickx, M. Pittman, and T. G. Kaye. "Oldest Preserved Umbilical Scar Reveals Dinosaurs Had "Belly Buttons." *BMC Biology* 20 (2022): 132.

Hunt, Katie. "This Fossil Reveals How Dinosaurs Peed, Pooped and Had Sex." CNN, January 19, 2021, https://www.cnn.com/2021/01/19/world/dinosaur-fossil-sex-study-scn/index .html.

Vinther, J., R. Nicholls, S. Lautenschlager, et al. "3D Camouflage in an Ornithischian Dinosaur." *Current Biology* 26 (2016): 2456–62.

Vinther, J., R. Nicholls, and D. A. Kelly. "A Cloacal Opening in a Non-avian Dinosaur." *Current Biology* 31 (2021): R1–R3.

Eighteen and Counting

Blackburn, D. G., and C. A. Sidor. "Evolution of Viviparous Reproduction in Paleozoic and Mesozoic Reptiles." *International Journal of Developmental Biology* 58 (2014): 935–48.

Caldwell, M. W., and M. S. Y. Lee. "Live Birth in Cretaceous Marine Lizards (Mosasauroids)." *Proceedings of the Royal Society B* 268 (2021): 2397–2401.

Cheng, Y.-n., X.-c. Wu, and Q. Ji. "Triassic Marine Reptiles Gave Birth to Live Young." *Nature* 432 (2004): 383–86.

Gao, K-Q., and D. T. Ksepka. "Osteology and Taxonomic Revision of *Hyphalosaurus* (Diapsida: Choristodera) from the Lower Cretaceous of Liaoning, China." *Journal of Anatomy* 212 (2008): 747–68.

Hou, L-H., P-P. Li, D. T. Ksepka, K-Q. Gao, and M. A. Norrel. "Implications of Flexible-Shelled Eggs in a Cretaceous Choristoderan Reptile." *Proceedings of the Royal Society B* 277 (2010): 1235–39.

Ji, Q., X-c. Wu, and Y-n. Cheng. "Cretaceous Choristoderan Reptiles Gave Birth to Live Young." *Naturwissenschaften* 97 (2010): 423–428.

Laird, M. K., M. B. Thompson, and C. M. Whittington. "Facultative Oviparity in a Viviparous Skink (*Saiphos equalis*)." *Biology Letters* 15 (2019): 20180827.

Liu, J, C. L. Organ, M. J. Benton, M. C. Brandley, and J. C. Aitchison. "Live Birth in an Archosauromorph Reptile." *Nature Communications* 8 (2017): article 14445.

Lü, J., Y. Kobayashi, C. D. Deeming, and Y. Liu. "Post-natal Parental Care in a Cretaceous Diapsid from Northeastern China." *Geosciences Journal* 19 (2015): 273–80.

Wang, M., L. Xing, K. Niu, Q. Liang, and S. Evans. "A New Specimen of the Early Cretaceous Long-Necked Choristodere *Hyphalosaurus* from Liaoning, China with Exceptionally-Preserved Integument." *Cretaceous Research* 144 (2023): 105451.

The "Lizard Fish" Moms

Maxwell, E. E., T. Argyriou, R. Stockar, and H. Furrer. "Re-evaluation of the Ontogeny and Reproductive Biology of the Triassic Fish *Saurichthys* (Actinopterygii, Saurichthyidae)." *Palaeontology* 61 (2018): 559–74.

Renesto, S., and R. Stockar. "Exceptional Preservation of Embryos in the Actinopterygian Saurichthys from the Middle Triassic of Monte San Giorgio, Switzerland." *Swiss Journal of Geosciences* 102 (2009): 323–30.

Whittington, C. M., and C. R. Friesen. "The Evolution and Physiology of Male Pregnancy in Syngnathid Fishes." *Biological Reviews* 95 (2020): 1252–72.

2. EGGS AND BABIES

Introduction

Salomon, M., J. Schneider, and Y. Lubin. "Maternal Investment in a Spider with Suicidal Maternal Care, *Stegodyphus lineatus* (Araneae, Eresidae)." *OIKOS* 109 (2005): 614–22.

Tethered Toddlers

Briggs, D. E. G., D. J. Siveter, D. J. Siveter, M. D. Sutton, and D. Legg. "Tiny Individuals Attached to a New Silurian Arthropod Suggest a Unique Mode of Brood Care." *PNAS* 113 (2016): 4410–15.

Briggs, D. E. G., D. J. Siveter, D. J. Siveter, M. D. Sutton, and D. Legg. "*Aquilonifer*'s Kites Are Not Mites." *PNAS* 113 (2016): E3320–21.

Piper, R. "Offspring or Phoronts? An Alternative Interpretation of the "Kite-Runner" Fossil." *PNAS* 113 (2016): E3319.

Vogt, G., and L. Tolley. "Brood Care in Freshwater Crayfish and Relationship with the Offspring's Sensory Deficiencies." *Journal of Morphology* 262 (2004): 566–82.

The Croc Guardian

Combrink, X., J. K. Warner, and C. T. Downs. "Nest Predation and Maternal Care in the Nile Crocodile (*Crocodylus niloticus*) at Lake St. Lucia, South Africa." *Behavioural Processes* 133 (2016): 31–36.

Hastings, A. K., and M. Hellmund. "Rare In Situ Preservation of Adult Crocodylian with Eggs from the Middle Eocene of Geiseltal, Germany." *PALAIOS* 30 (2015): 446–61.

Ammonite Eggs

Etches, S., J. Clarke, and J. Callomon. "Ammonite Eggs and Ammonitellae from the Kimmeridge Clay Formation (Upper Jurassic) of Dorset, England." *Lethaia* 42 (2009): 204–17.

Lomax, D. R., and B. G. Hyde. "Ammonite Aptychi from the Lower Jurassic (Toarcian) Near Whitby, North Yorkshire, UK." *Proceedings of the Yorkshire Geological Society* 59 (2012): 99–107.

Zatoń, M., and A. A. Mironenko. "Exceptionally Preserved Late Jurassic Gastropod Egg Capsules." *PALAIOS* 30 (2015): 482–89.

That's a Lot of Babies!

Hoffman, E. A., and T. B. Rowe. "Jurassic Stem-Mammal Perinates and the Origin of Mammalian Reproduction and Growth." *Nature* 561 (2018): 104–8.

Sues, H-D. "Palaeontology: Many Babies or Bigger Brains?" *Current Biology* 28 (2018): R1243–65.

3. FAMILY AND FRIENDS

Introduction

Hiemstra, A-F., C. W. Moeliker, B. Gravendeel, and M. Schilthuizen. "Bird Nests Made from Anti-bird Spikes." *Deinsea* 21 (2023): 17–25.

The Mother in the Tree

Botha-Brink, J., and S. P. Modesto. "A mixed-age classed 'pelycosaur' aggregation from South Africa: earliest evidence of parental care in amniotes?" *Proceedings of the Royal Society B* 274 (2007): 2829–34.

Jasinoski, S. C., and F. Abdala. "Aggregations and Parental Care in the Early Triassic Basal Cynodonts *Galesaurus planiceps* and *Thrinaxodon liorhinus*." *PeerJ* 5 (2017): e2875.

Maddin, H. C., A. Mann, and B. Hebert. "Varanopid from the Carboniferous of Nova Scotia Reveals Evidence of Parental Care in Amniotes." *Nature Ecology & Evolution* 4 (2020): 50–56.

Vickaryous, M. K. and J-Y. Sire. "The Integumentary Skeleton of Tetrapods: Origin, Evolution, and Development." *Journal of Anatomy* 214 (2009): 441–64.

What You (and I) Didn't See at Egg Mountain

DeMar, D. G., J. L. Conrad, J. J. Head, D. J. Varricchio, and G. P. Wilson "A New Late Cretaceous Iguanomorph from North America and the Origin of New World Pleurodonta (Squamata, Iguania)." *Proceedings of the Royal Society B* 284 (2016): 20161902.

Horner, J. R., and R. Makela. "Nest of Juveniles Provides Evidence of Family Structure Among Dinosaurs." *Nature* 282 (1979): 296–98.

Weaver, L. N., D. J. Varricchio, E. J. Sargis, et al. "Early Mammalian Social Behaviour Revealed by Multituberculates from a Dinosaur Nesting Site." *Nature Ecology & Evolution* 5 (2020): 32–37.

The Watery Graveyard

Chiba, K., M. J. Ryan, D. R. Braman, et al. "Taphonomy of a Monodominant *Centrosaurus apertus* (Dinosauria: Ceratopsia) Bonebed from the Upper Oldman Formation of Southeastern Alberta." *PALAIOS* 30 (2015): 655–67.

Eberth, D. A. *A Review of Ceratopsian Paleoenvironmental Associations and Taphonomy: New Perspectives on Horned Dinosaurs*. Indiana University Press, 2010.

Eberth, D. A. "Origins of Dinosaur Bonebeds in the Cretaceous of Alberta, Canada." *Canadian Journal of Earth Sciences* 52 (2015): 655–81.

Eberth, D. A., and M. A. Getty "Ceratopsian Bonebeds: Occurrences, Origins and Significance." In *Dinosaur Provincial Park: A Spectacular Ancient Ecosystem Revealed*. Indiana University Press, 2005.

Eberth, D. A., D. B. Brinkman, and V. A. Barkas. "Centrosaurine Mega-Bonebed from the Upper Cretaceous of Southern Alberta: Implications for Behavior and Death Events." In *New Perspectives on Horned Dinosaurs*. Indiana University Press, 2010.

Hunt, R. K., and A. A. Farke. "Behavioral Interpretations from Ceratopsid Bonebeds." In *New Perspectives on Horned Dinosaurs*. Indiana University Press, 2010.

Ryan, M. J., A. P. Russell, D. A. Eberth, and P. J. Currie. "The Taphonomy of a *Centrosaurus* (Ornithischia: Certopsidae) Bone Bed from the Dinosaur Park Formation (Upper Campanian), Alberta, Canada, with Comments on Cranial Ontogeny." *PALAIOS* 16 (2001): 482–506.

Armor and Out

Ahlberg, P. E., and Z. Johanson. "Second Tristichopterid (Sarcopterygii, Osteolepiformes) from the Upper Devonian of *Canowindra*, New South Wales, Australia, and Phylogeny of the Tristichopteridae." *Journal of Vertebrate Paleontology* 17, no. 4 (1997): 653–73.

Ritchie, A. "The Great Devonian Fish Kill at Canowindra." In *Evolution and Biogeography of Australasian Vertebrates*, ed. J. R. Merrick, M. Archer, G. M. Hickey, and M. S. Y. Lee. Auscipub, 2006.

Ritchie, A. and Johanson, Z. "Buried Treasures from the Age of Fishes." *Australian Age of Dinosaurs* 4 (2006): 1–45.

4. MOVING ALONG

Introduction

Bressman, N. R., J. E. Hill, and M. A. Ashley-Ross. "Why Did the Invasive Walking Catfish Cross the Road? Terrestrial Chemoreception Described for the First Time in a Fish." *Journal of Fish Biology* 97 (2020): 895–907.

Bressman, N. R., C. H. Morrison, and M. A. Ashley-Ross. "Reffling: A Novel Locomotor Behavior Used by Neotropical Armored Catfishes (Loricariidae) in Terrestrial Environments." *Ichthyology & Herpetology* 109 (2021): 608–25.

Swimming Dinosaurs

Milner, A. R. C., M. G. Lockley, and J. I. Kirkland. "A Large Collection of Well-Preserved Theropod Dinosaur Swim Tracks from the Lower Jurassic Moenave Formation, St. George, Utah." *New Mexico Museum of Natural History and Science Bulletin* 37 (2006): 315–28.

Milner, A. R. C., and M. G. Lockley. "Dinosaur Swim Track Assemblages: Characteristics, Contexts, and Ichnofacies Implications." In *Dinosaur Tracks*, Indiana University Press, 2016.

Whyte, M. A., and M. Romano. "A Dinosaur Ichnocoenosis from the Middle Jurassic of Yorkshire, UK." *Ichnos* 8 (2021): 223–34.

The Death Star

Baumiller, T. K., and C. G. Messing. "Stalked Crinoid Locomotion, and Its Ecological and Evolutionary Implications." *Palaeontologia* Electronica (2007): 10, 1–10.

Brom, K. R., K. Oguri, T. Oji, M. A. Salamon, and P. Gorzelak. "Experimental Neoichnology of Crawling Stalked Crinoids." *Swiss Journal of Palaeontology* 137 (2018): 197–203.

De Carvalho, C. N., B. Pereira, A. Klompmaker, et al. "Running Crabs, Walking Crinoids, Grazing Gastropods: Behavioral Diversity and Evolutionary Implications of the Cabeço da Ladeira Lagerstätte (Middle Jurassic, Portugal)." *Comunicações Geológicas* 103 (2016): 39–54.

Myers, R. A., C. M. Furlong, M. K. Gingras, and J-P. Zonneveld. "Locomotion Traces Emplaced by Modern Stalkless Comatulid Crinoids (Featherstars)." *PALAIOS* 38 (2023): 474–89.

Star Kisses

Brown, B., and H. E. Vokes. "Fossil Imprints of Unknown Origin: Further Information and a Possible Explanation." *American Journal of Science* 242 (1944): 656–72.

Connely, M. V. 2019. Vertebrate trace fossils in the Mowry Shale (lower cretaceous) of Wyoming, USA. Paludicola, 12, 68–82.

Donovan, D. T., and D. Fuchs. "Part M, Chapter 13: Fossilized Soft Tissues in Coleoidea." *Treatise Online* 73 (2016): 1–30.

Fuchs, D., D. T. Donovan, and H. Keupp. "Taxonomic Revision of *"Onychoteuthis"* conocauda Quenstedt, 1849 (Cephalopoda: Coleoidea)." *Neues Jahrbuch für Geologie und Paläontologie* 270, no. 3 (2013): 245–55.

Vokes, H. E. "Fossil Imprints of Unknown Origin." *American Journal of Science* 239 (1941): 451–53.

Hitching a Ride in Style

Penney, D., A. McNeil, D. I. Green, et al. "Ancient Ephemeroptera–Collembola Symbiosis Fossilized in Amber Predicts Contemporary Phoretic Associations." *PLOS One* 7 (2012): e47651.

Robin, N., C. D'Haese, and P. Barden. "Fossil Amber Reveals Springtails' Longstanding Dispersal by Social Insects." *BMC Evolutionary Biology* 19 (2019): 213.

Watch Where You Step

Halaçlar, K., P. Rummy, T. Deng, and V. T. Do. "Footprint on a Coprolite: A Rarity from the Eocene of Vietnam." *Palaeoworld* 31 (2022): 723–32.

Halaçlar, K., P. Rummy, J. Liu, et al. "Exceptionally Well-Preserved Crocodilian Coprolites from the Late Eocene of Northern Vietnam: Ichnology and Paleoecological Significance." *iScience* 26 (2023): 107607.

5. HIDE-AND-SEEK

Introduction

Brandt, M., and D. Mahsberg. "Bugs with a Backpack: The Function of Nymphal Camouflage in the West African Assassin Bugs *Paredocla* and *Acanthaspis* spp." *Animal Behaviour* 63 (2002): 277–84.

Autotomy for the Win

Barr, J. I., R. Somaweera, S. S. Godfrey, M. G. Gardner, and P. W. Bateman. "When One Tail Isn't Enough: Abnormal Caudal Regeneration in Lepidosaurs and Its Potential Ecological Impacts." *Biological Reviews* 95 (2020): 1479–96.

LeBlanc, A. R. H., M. J. MacDougall, Y. Haridy, D. Scott, and R. R. Reisz. "Caudal Autotomy as Anti-predatory Behavior in Paleozoic Reptiles." *Scientific Reports* 8 (2017): 3328.

Lozito, T. P., and R. S. Tuan. "Lizard Tail Regeneration as an Instructive Model of Enhanced Healing Capabilities in an Adult Amniote." *Connective Tissue Research* 58 (2017): 145–54.

Simões, T. R., M. W. Caldwell, R. L. Nydam, and P. Jiménez-Huidobro. "Osteology, Phylogeny, and Functional Morphology of Two Jurassic Lizard Species and the Early Evolution of Scansoriality in Geckoes." *Zoological Journal of the Linnean Society* 180 (2017): 216–41.

Xu, C., J. Palade, R. E. Fisher, et al. "Anatomical and Histological Analyses Reveal That Tail Repair Is Coupled with Regrowth in Wild-Caught, Juvenile American Alligators (*Alligator mississippiensis*)." *Scientific Reports* 10 (2020): 20122.

Decapod Housing Supplies

Fraaije, E. H. B. "The Oldest *In Situ* Hermit Crab from the Lower Cretaceous of Speeton, UK." *Palaeontology* 46 (2023): 55–57.

Fraaije, E. H. B., J. W. M. Jagt, B. W. M van Bakel, and D. M. Tshudy. "A New Early Late Cretaceous Nephropid Lobster (Crustacea, Decapoda) from Kazakhstan, Entombed Within an Ammonite Body Chamber." *Cretaceous Research* 115 (2020): 104552.

Fraaye, R. H. B., and M. Jager. "Decapods in Ammonite Shells: Examples of Inquilinism from the Jurassic of England and Germany. *Palaeontology* 38 (1995): 63–75.

Hyžný, M., and A. Klompmaker. "Systematics, Phylogeny, and Taphonomy of Ghost Shrimps (Decapoda): A Perspective from the Fossil Record." *Arthropod Systematics & Phylogeny* 73 (2015): 401–37.

Jagt, J. W. M., B. W. M. van Bakel, R. H. B. Fraaije, and C. Neumann. "In Situ Fossil Hermit Crabs (Paguroidea) from Northwest Europe and Russia: Preliminary Data on New Records." *Revista Mexicana de Ciencias Geológicas* 23 (2006): 364–69.

Jenkins, R. J. F. "The Fossil Crab *Ommatocarcinus corioensis* (Cresswell) and a Review of Related Australian Species." *Memoirs of Museum Victoria* 36 (1975): 33–62.

Klompmaker, A. A., and R. H. B. Fraaije. "Animal Behavior Frozen in Time: Gregarious Behavior of Early Jurassic Lobsters Within an Ammonoid Body Chamber." *PLOS One* 7 (2012): 1–9.

Landman, N. H., R. H. B. Fraaije, S. M. Klofak, et al. "Inquilinism of a Baculite by a Dynomenid Crab from the Upper Cretaceous of South Dakota." *American Museum Novitates* 3818 (2014): 1–16.

Mironenko, A. "A Hermit Crab Preserved Inside an Ammonite Shell from the Upper Jurassic of Central Russia: Implications to Ammonoid Palaeoecology." *Palaeogeography, Palaeoclimatology, Palaeoecology* 537 (2020): 109397.

Schweigert, G., R. Fraaije, P. Havlik, and A. Nützel. "New Early Jurassic Hermit Crabs from Germany and France." *Journal of Crustacean Biology* 33 (2013): 802–17.

Stilwell, J. D., R. H. Levy, R. M. Feldmann, and D. M. Harwood. "On the Rare Occurrence of Eocene Callianassid Decapods (Arthropoda) Preserved in Their Burrows, Mount Discovery, East Antarctica." *Journal of Paleontology* 71 (1997): 284–87.

van Bakel, B. W. M., R. H. B Fraaije, J. W. M. Jagt, and P. Artal. "An Unexpected Diversity of Late Jurassic Hermit Crabs (Crustacea, Decapoda, Anomura) in Central Europe." *Neues Jahrbuch für Geologie und Paläontologie* 250 (2008): 137–56.

Walker, S. E. "Biological Remanie; Gastropod Fossils Used by the Living Terrestrial Hermit Crab, *Coenobita clypeatus*, on Bermuda." *PALAIOS* 9 (1994): 403–12.

Tiny Homes for Tiny Amphibians

Hembree, D. I., L. D. Martin, and S. T. Hasiotis. "Amphibian Burrows and Ephemeral Ponds of the Lower Permian Speiser Shale, Kansas: Evidence for Seasonality in the Mid-Continent." *Palaeogeography, Palaeoclimatology, Palaeoecology* 203 (2004): 127–52.

Hembree, D. I., S. T. Hasiotis, and L. D. Martin. "*Torridorefugium eskridgensis* (New Ichnogenus and Ichnospecies): Amphibian Aestivation Burrows from the Lower Permian Speiser Shale of Kansas." *Journal of Paleontology* 79 (2005): 596–606.

Pardo, J. D., and J. S. Anderson. "Cranial Morphology of the Carboniferous-Permian Tetrapod *Brachydectes newberryi* (Lepospondyli, Lysorophia): New Data from μCT." *PLOS One* 11 (2016): e0161823.

Undercover Insect

Garrouste, R., S. Hugel, L. Jacquelin, et al. "Insect Mimicry of Plants Dates Back to the Permian." *Nature Communications* 7 (2016): 13735.

Ghirotto, V. M., E. B. Crispino, P. I. Chiquetto-Machado, et al. "The Oldest Euphasmatodea (Insecta, Phasmatodea): Modern Morphology in an Early Cretaceous Stick Insect Fossil from the Crato Formation of Brazil." *Papers in Palaeontology* 8 (2022): e1437.

Jouault, C., H. Tischlinger, M. Henrotay, and A. Nel. "Wing Coloration Patterns in the Early Jurassic Dragonflies as Potential Indicator of Increasing Predation Pressure from Insectivorous Reptiles." *Palaeoentomology* 5 (2022): 305–18.

Logghe, A., A. Nel, J-S. Steyer, et al. "A Twig-Like Insect Stuck in the Permian Mud Indicates Early Origin of an Ecological Strategy in Hexapoda Evolution." *Scientific Reports* 11 (2021): 20774.

Schultz, M. *Mimicry in Insects: An Illustrated Study in Mimicry and Cryptic Coloration in Insects.* University of Nebraska–Lincoln, 2018.

Wang, Y-J., Z-Q. Liu, X. Wang, et al. "Ancient Pinnate Leaf Mimesis Among Lacewings." *PNAS* 107 (2010): 16212–15.

Wang, Y-J., C. C. Labandeira, C-K. Shih, et al. "Jurassic Mimicry Between a Hangingfly and a Ginkgo from China." *PNAS* 109 (2012): 20514–19.

Wedman, S. "A Brief Review of the Fossil History of Plant Masquerade by Insects." *Palaeontographica Abteilung B* 283 (2010): 175–82.

6. FINDING FOOD

Introduction

Deban, S. M., J. C. O'Reilly, U. Dicke, and J. L. van Leeuwen. "Extremely High-Power Tongue Projection in Plethodontid Salamanders." *Journal of Experimental Biology* 210 (2007): 655–67.

Noel, A. C., and D. L. Hu. "The Tongue as a Gripper." *Journal of Experimental Biology* 221 (2018): jeb176289.

I Got Your Chicken Legs

Therrien, F., D. K. Zelenitsky, K. Tanaka, et al. "Exceptionally Preserved Stomach Contents of a Young Tyrannosaurid Reveal an Ontogenetic Dietary Shift in an Iconic Extinct Predator." *Science Advances* 9 (2023): eadi0505.

Varricchio, D. J. "Gut Contents from a Cretaceous Tyrannosaurid: Implications for Theropod Dinosaur Digestive Tracts." *Journal of Paleontology* 75 (2011): 401–6.

A Not So Whale of a Time

Fahlke, J. M. "Bite Marks Revisited–Evidence for Middle-to-Late Eocene *Basilosaurus isis* Predation on *Dorudon atrox* (Both Cetacea, Basilosauridae)." *Palaeontologia Electronica* 15 (2012): 1–16.

Gingerich, P. D. "Wadi Al-Hitan or 'Valley of Whales'—an Eocene World Heritage Site in the Western Desert of Egypt." *Geological Society Special Publication* 543 (2023): 421–30.

Snively, E., J. M. Fahlke, and R. C. Welsh. "Bone-Breaking Bite Force of *Basilosaurus isis* (Mammalia, Cetacea) from the Late Eocene of Egypt Estimated by Finite Element Analysis." *PLOS One* 10 (2015): e0118380.

Swift, C. C., and L. G. Barnes. "Stomach Contents of *Basilosaurus cetoides*: Implications for the Evolution of Cetacean Feeding Behavior, and the Evidence for Vertebrate Fauna of Epicontinental Eocene Seas." *Paleontological Society Special Publications* 8 (1996): 380.

Totterdell, J. A., R. Wellard, I. M. Reeves, et al. "The First Three Records of Killer Whales (*Orcinus orca*) Killing and Eating Blue Whales (*Balaenoptera musculus*)." *Marine Mammal Science* 38 (2022): 1286–1301.

Voss, M., M. S. M. Antar, I. S. Zalmout, and P. D. Gingerich. "Stomach Contents of the Archaeocete *Basilosaurus*: Apex Predator in Oceans of the Late Eocene." *PLOS One* 14 (2019): e0209021.

The Early Bird Gulps the Fruit

Hu, H., Y. Wang, Y., P. G. McDonald, et al. "Earliest Evidence for Fruit Consumption and Potential Seed Dispersal by Birds." *eLife* 11 (2022): e74751.

O'Connor, J. K., A. Clark, F. Herrera, et al. "Direct Evidence of Frugivory in the Mesozoic Bird *Longipteryx* Contradicts Morphological Proxies for Diet." *Current Biology* 34 (2024): 4559–66.

Wu, Y., Y. Ge, H. Hu, et al. "Intra-gastric Phytoliths Provide Evidence for Folivory in Basal Avialans of the Early Cretaceous Jehol Biota." *Nature Communications* 14 (2023): 4558.

Zheng, X., L. D. Martin, Z. Zhou, et al. "Fossil Evidence of Avian Crops from the Early Cretaceous of China." *PNAS* 108 (2011): 15904–7.

Zhou, Z., and F. Zhang. "A Long-Tailed, Seed-Eating Bird from the Early Cretaceous of China." *Nature* 418 (2002): 405–9.

When the Tables Turn

Hart, M. B., G. Arratia, C. Moore, and B. J. Ciotti. "Life and Death in the Jurassic Seas of Dorset, Southern England." *Proceedings of the Geologists' Association* 131 (2020): 629–38.

Jenny, D., D. Fuchs, A. I. Arkhipkin, et al. "Predatory Behavior and Taphonomy of a Jurassic Belemnoid Coleoid (Diplobelida, Cephalopoda)." *Scientific Reports* 9 (2019): 1–11.

Klug, C., G. Schweigert, R. Hoffmann, R. Weis, and K. De Baets. "Fossilized Leftover Falls as Sources of Palaeoecological Data: A 'Pabulite' Comprising a Crustacean, a Belemnite and a Vertebrate from the Early Jurassic Posidonia Shale." *Swiss Journal of Palaeontology* 140 (2021): 10.

7. CONFLICT

Introduction

Earth Touch News. "Dramatic Footage Shows Snow Leopard Plummeting Down a Cliff While Clutching Its Prey." Earth Touch News Network, December 26, 2018, accessed February 11, 2025. https://www.earthtouchnews.com/natural-world/predator-vs-prey/dramatic-footage-shows-snow-leopard-plummeting-down-a-cliff-while-clutching-its-prey/.

Taco Takedown and the Eternal "Hug"

Han, G., J. C. Mallon, A. J. Lussier, et al. "An Extraordinary Fossil Captures the Struggle for Existence During the Mesozoic." *Scientific Reports* 13 (2023): 11221.

MacLennan, S.A., et al. "Extremely Rapid, Yet Noncatastrophic, Preservation of the Flattened-Feathered and 3D Dinosaurs of the Early Cretaceous of China." *PNAS* 121 (2024): e2322875121.

Four Daggers Better than Two

Boyd, C. L., E. Starck, W. Welsh, and M. Householder. "Bite Marks on Nimravid Crania and Implications for Intraclade Interactions Within Nimravidae (Mammalia: Feliformia)." *GSA Annual Meeting* 45 (2013): 755.

Brown, J. G. "Jaw Function in *Smilodon fatalis*: A Reevaluation of the Canine Shear-Bite and a Proposal for a New Forelimb-Powered Class 1 Lever Model." *PLOS One* 9 (2014): e107456

Chimento, N. R., F. L. Agnolin, L. Soibelzon, J. G. Ochoa, and V. Buide. V. "Evidence of Intraspecific Agonistic Interactions in *Smilodon populator* (Carnivora, Felidae)." *Comptes Rendus Palevol* 18 (2019): 449–54.

Geraads, D., T. Kaya, and V. Tuna. "A Skull of *Machairodus giganteus* (Felidae, Mammalia) from the Late Miocene of Turkey." *Neues Jahrbuch für Geologie und Paläontologie* 2 (2004): 95–110.

Reynolds, A. R., K. L. Seymour, and D. C. Evans. "*Smilodon fatalis* Siblings Reveal Life History in a Sabertoothed Cat." *iScience* 24 (2020): 101916.

Croc Dino Dinner

Blanco, R. E., W. W. Jones, and J. Villamil. "The 'Death Roll' of Giant Fossil Crocodyliforms (Crocodylomorpha: Neosuchia): Allometric and Skull Strength Analysis." *Historical Biology* 27 (2014): 514–24.

Cossette, A. P., and C. A. Brochu. "A Systematic Review of the Giant Alligatoroid *Deinosuchus* from the Campanian of North America and Its Implications for the Relationships at the Root of Crocodylia." *Journal of Vertebrate Paleontology* 40, no. 1 (2020): e1767638.

Rivera-Sylva, H. E., E. Frey, and J. R. Guzmán-Gutiérrez. "Evidence of Predation on the Vertebra of a Hadrosaurid Dinosaur from the Upper Cretaceous (Campanian) of Coahuila, Mexico." *Notebooks on Geology* CG2009 L02 (2009).

Schwimmer, D. R. *King of the Crocodylians: The Paleobiology of* Deinosuchus. Indiana University Press, 2002.

Schwimmer, D. R. "Bite Marks of the Giant Crocodilian *Deinosuchus* on Late Cretaceous (Campanian) Bones." In *Crocodyle Tracks and Traces. New Mexico Museum of Natural History and Science Bulletin* 51 (2010): 183–90.

Tennant, J. P. "Fossil Focus: Mesozoic Crocodyliforms." *Palaeontology Online* 6 (2016): 1–15.

White, M. A., P. R. Bell, N. E. Campione, et al. "Abdominal Contents Reveal Cretaceous Crocodyliforms Ate Dinosaurs." *Gondwana Research* 106 (2022): 281–302.

Smash That

Arbour, V. M. "Estimating Impact Forces of Tail Club Strikes by Ankylosaurid Dinosaurs." *PLOS One* 4 (2009): e6738.

Arbour, V. M., and P. J. Currie. "Tail and Pelvis Pathologies of Ankylosaurian Dinosaurs." *Historical Biology* 23 (2011): 375–90.

Arbour, V. M., and D. C. Evans. "A New Ankylosaurine Dinosaur from the Judith River Formation of Montana, USA, Based on an Exceptional Skeleton with Soft Tissue Preservation." *Royal Society Open Science* 4 (2017): 161086.

Arbour, V. M., L. E. Zanno, and D. C. Evans. "Palaeopathological Evidence for Intraspecific Combat in Ankylosaurid Dinosaurs." *Biology Letters* 18 (2022): 20220404.

Park, J. Y., Y. N. Lee, Y. Kobayashi, et al. "A New Ankylosaurid from the Upper Cretaceous Nemegt Formation of Mongolia and Implications for Paleoecology of Armoured Dinosaurs." *Scientific Reports* 11 (2021): 22928.

Clash of the Mighty Marine Lizards

Everhart, M. J. "A Bitten Skull of *Tylosaurus kansasensis* (Squamata: Mosasauridae) and a Review of Mosasaur-on-Mosasaur Pathology in the Fossil Record." *Transactions of the Kansas Academy of Science* 111 (2008): 251–62.

Jiménez-Huidobro, P., T. R. Simões, and M. W. Caldwell. "Re-characterization of *Tylosaurus nepaeolicus* (Cope, 1874) and *Tylosaurus kansasensis* (Everhart, 2005): Ontogeny or Sympatry?" *Cretaceous Research* 65 (2016): 68e81.

8. DIFFERENT DIETS

Introduction

Pierce, R. "Chickens Eating Mice: Is It Dangerous and How Can You Limit Mouse Visits?" Morning Chores blog, 2024. Accessed February 11, 2025. https://morningchores.com/chickens-eating-mice/.

Mosasaur Mash Up

Bell, G. L., Jr. and J. E. Martin. "Direct Evidence of Aggressive Intraspecific Competition in *Mosasaurus conodon* (Mosasauridae: Squamata)." *Journal of Vertebrate Paleontology* 15 (1995): 18A.

Everhart, M. J. "A Bitten Skull of *Tylosaurus kansasensis* (Squamata: Mosasauridae) and a Review of Mosasaur-on-Mosasaur Pathology in the Fossil Record." *Transactions of the Kansas Academy of Science* 111 (2008): 251–62.

Konishi, T., and M. W. Caldwell. "Two New Plioplatecarpine (Squamata, Mosasauridae) Genera from the Upper Cretaceous of North America, and a Global Phylogenetic Analysis of plioplatecarpines." *Journal of Vertebrate Paleontology* 31 (2011): 754–83.

Konishi, T. "Non-lethal Face Biting Between Mosasaurs (Squamata: Mosasauridae): The First Unequivocal Evidence from an Exceptional Skeleton of *Mosasaurus* sp. from Southern Alberta, Canada." Society of Vertebrate Paleontology abstract, 2016.

Polcyn, M. J., A. S. Schulp, and A. O. Goncalves. "Remarkably Well-Preserved In-Situ Gut-Content in a Specimen of *Prognathodon kianda* (Squamata: Mosasauriae) Reveals Multispecies Intrafamilial Predation, Cannibalism, and a New Mosasaurine Taxon: Windows into Sauropsid and Synapsid Evolution." Dinosaur Science Center Press, 2023.

Tyskoski, R. S., and M. J. Polcyn. "'Tis but a scratch!" Said the Black Knight; Severe Facial Pathologies in a *Tylosaurus* from the Ozan Formation (Campanian) of Northeast Texas." Society of Vertebrate Paleontology abstract, 2023.

Micro Meals—Food for Thought

Hone, D. W. E., T. A. Dececchi, C. Sullivan, X. Xing, and H. C. E. Larsson. "Generalist Diet of *Microraptor zhaoianus* Included Mammals." *Journal of Vertebrate Paleontology* 42, no. 2 (2022): e2144337.

O'Connor, J., Z. Zhou, and X. Xu. "Additional Specimen of *Microraptor* Provides Unique Evidence of Dinosaurs Preying on Birds." *Proceedings of the National Academy of Sciences* 108 (2011): 19662–65.

O'Connor, J., X. Zheng, L. Dong, et al. "*Microraptor* with Ingested Lizard Suggests Nonspecialized Digestive Function." *Current Biology* 29 (2019): 2423–29.

Xing, L., W. S. Persons IV, P. R. Bell, et al. "Piscivory in the Feathered Dinosaur *Microraptor*." *Evolution* 67 (2013): 2441–45.

Xing, L., P. R. Bell, P. R., W. S. Persons IV, et al. "Abdominal Contents from Two Large Early Cretaceous Compsognathids (Dinosauria: Theropoda) Demonstrate Feeding on Confuciusornithids and Dromaeosaurids." *PLOS One* 7 (2012): e44012.

Xu, X., Z. Zhou, X. Wang, et al. "Four-Winged Dinosaurs from China." *Nature* 421 (2003): 335–40.

Fish Are Food, Not Friends

Buffetaut, E., D. Martill, and F. Escuillié. "Pterosaurs as Part of a Spinosaur Diet." *Nature* 430 (2004): 33.

Charig, A. J., and A. C. Milner. "*Baryonyx*, a Remarkable New Theropod Dinosaur." *Nature* 324 (1986): 359–61.

Charig, A. J., and A. C. Milner. "*Baryonyx walkeri*, a Fish-Eating Dinosaur from the Wealden of Surrey. *Bulletin of the Natural History Museum, Geology Series* 53 (1997): 11–70.

Lomax, D. R., and N. Tamura. *Dinosaurs of the British Isles*. Siri Scientific Press, 2014.

Sea Dragon Supper Surprise

Böttcher, R. "Uber die Nhrung eines *Leptopterygius* (Ichthyosauria, Reptilia) aus dem sfddeutschen Posidonienschiefer (Unterer Jura) mit Bemerkungen fiber den Magen der Ichthyosaurier." *Stuttgarter Beitrdge Naturkunde B* 155 (1989): 1–19.

Dick, D. G., G. Schweigert, and E. E. Maxwell. "Trophic Niche Ontogeny and Palaeoecology of Early Toarcian *Stenopterygius* (Reptilia: Ichthyosauria)." *Palaeontology* 59 (2016): 423–31.

Jiang, D-Y., R. Motani, A. Tintori, et al. "Evidence Supporting Predation of 4-m Marine Reptile by Triassic Megapredator." *iScience* 23 (2020): 101347.

Kear, B. P., W. E. Boles, and E. T. Smith. "Unusual Gut Contents in a Cretaceous Ichthyosaur." *Proceedings of the Royal Society of London B* 270 (2003): 206–8.

Lomax, D. R. "An *Ichthyosaurus* (Reptilia, Ichthyosauria) with Gastric Contents from Charmouth, England: First Report of the Genus from the Pliensbachian." *Paludicola* 8 (2010): 22–36.

Lomax, D. R., and J. A. Massare. "A New Species of *Ichthyosaurus* from the Lower Jurassic of West Dorset, England." *Journal of Vertebrate Paleontology* 35 (2015): 1–14.

Pollard, J. E. "The Gastric Contents of an Ichthyosaur from the Lower Lias of Lyme Regis." *Palaeontology* 11 (1968): 376–88.

9. DIGESTION

Introduction

Brunnschweiler, J. M., P. L. R. Andrews, E. J. Southall, M. Pickering, and D. W. Sims. "Rapid Voluntary Stomach Eversion in a Free-Living Shark. *Journal of the Marine Biological Association of the UK* 85 (2005): 1141–44.

Buckland, W. "On the Discovery of Coprolites, or Fossil Faeces, in the Lias at Lyme Regis, and in Other Formations." *Transactions of the Geological Society of London* 3 (1829): 223–236.

A Coprolite Fit for a King . . . or Maybe a Queen

Chin, K., T. T. Tokaryk, G. M. Erickson, and L. C. Calk. "A King-Sized Theropod Coprolite." *Nature* 393 (1998): 680–82.

Chin, K., D. A. Eberth, M. H. Schweitzer, et al. "Remarkable Preservation of Undigested Muscle Tissue Within a Late Cretaceous Tyrannosaurid Coprolite from Alberta, Canada." *PALAIOS* 18 (2003): 286–94.

Guinness World Records. "Largest Coprolite from a Carnivore," 2020. Accessed February 11, 2025. https://www.guinnessworldrecords.com/world-records/78871-largest-coprolite-fossilised-excrement-from-a-carnivore.

The Scoop on Some Rather Unusual Poop

Chin, K., R. M. Feldman, and J. N. Tashman. "Consumption of Crustaceans by Megaherbivorous Dinosaurs: Dietary Flexibility and Dinosaur Life History Strategies." *Scientific Reports* 7 (2017): 11163.

Dentzien-Dias, P. C., G. Poinar, A. E. Q. de Figueiredo, et al. "Tapeworm Eggs in a 270-Million-Year-Old Shark Coprolite." *PLOS One* 8 (2013): e55007.

Qvarnström, M., J. V. Wernström, R. Piechowski, et al. "Beetle-Bearing Coprolites Possibly Reveal the Diet of a Late Triassic Dinosauriform." *Royal Society Open Science* 6 (2019): 181042.

Don't Forget to Eat Your Stones

Cerda, I. A. "Gastroliths in an Ornithopod Dinosaur." *Acta Palaeontologica Polonica* 53 (2008): 351–55.

Currie, P. J. "*Hovasaurus boulei*: An Aquatic Eosuchian from the Upper Permian of Madagascar." *Palaeontologica Africana* 24 (1981): 99–168.

Henderson, D. M. "Lost, Hidden, Broken, Cut-Estimating and Interpreting the Shapes and Masses of Damaged Assemblages of Plesiosaur Gastroliths." *PeerJ* 12 (2024): e17925.

Hunt, A. P., and S. P. Lucas. "The Ichnology of Vertebrate Consumption: Dentalites, Gastroliths and Bromalites." *New Mexico Museum of Natural History & Science Bulletin* 87 (2021): 1–216.

Kubo, T., M. T. Mitchell, and D. M. Henderson "*Albertonectes vanderveldei*, a New Elasmosaur (Reptilia, Sauropterygia) from the Upper Cretaceous of Alberta." *Journal of Vertebrate Paleontology* 32 (2012): 557–72.

Nudds, J. R., D. R. Lomax, and J. P. Tennant. "Gastroliths and Deinonychus Teeth Associated with a Skeleton of Tenontosaurus from the Cloverly Formation (Lower Cretaceous), Montana, USA." *Cretaceous Research* (2022): 105327.

O'Gorman, J. P., E. B. Oliveiro, S. Santillana, S., M. J. Everhart, and M. Reguero. "Gastroliths Associated with an *Aristonectes* Specimen (Plesiosauria, Elasmosauriodae), López de Bertodano Formation (Upper Maastrichtian) Seymour Island (Is. Marambio), Antarctic Peninsula." *Cretaceous Research* 50 (2014): 228–37.

Thompson, W. A., J. E. Martin, and M. Requero. "Comparison of Gastroliths Within Plesiosaurs (Elasmosauridae) from the Late Cretaceous Marine Deposits of Vega Island, Antarctic Peninsula and the Missouri River Area, South Dakota." *Geological Society of America Special Paper* 427 (2007): 146–53.

Uriona, T. J., M. Lyon, and C. G. Farmer. "Lithophagy Prolongs Voluntary Dives in American Alligators (*Alligator mississippiensis*)." *Integrative Organismal Biology* (2018): 1–4.

Wings, O. "The Rarity of Gastroliths in Sauropod Dinosaurs—A Case Study in the Late Jurassic Morrison Formation, Western USA." *Fossil Record* 18 (2015): 1–16.

Cough It Up!

Foster, J. R. A. P. Hunt, and J. I. Kirkland. "Significance of a Small Regurgitalite Containing Lissamphibian Bones, from the Morrison Formation (Upper Jurassic), Within a Diverse Plant Locality Deposit in Southeastern Utah, USA." *PALAIOS* 37 (2022): 433–42.

Freimuth, W. J., D. J. Varricchio, A. L. Brannick, L. N. Weaver, and G. Wilson Mantilla. "Mammal-Bearing Gastric Pellets Potentially Attributable to *Troodon formosus* at the Cretaceous Egg Mountain Locality, Two Medicine Formation, Montana, USA." *Palaeontology* 64 (2021): 699–725.

Hunt, A. P., and S. P. Lucas. "The Ichnology of Vertebrate Consumption: Dentalites, Gastroliths and Bromalites." *New Mexico Museum of Natural History & Science Bulletin* 87 (2021): 1–216.

Jiang, S., X. Wang, X. Zheng, et al. "Two Emetolite-Pterosaur Associations from the Late Jurassic of China: Showing the First Evidence for Antiperistalsis in Pterosaurs." *Philosophical Transactions of the Royal Society B* 377 (2022): 20210043.

Myhrvold, N. P. "A Call to Search for Fossilised Gastric Pellets." *Historical Biology* 24 (2012): 505–17.

Zheng, X., X. Wang, C. Sullivan, C., et al. "Exceptional Dinosaur Fossils Reveal Early Origin of Avian-Style Digestion." *Scientific Reports* 8 (2018): 14217.

10. HEALTH AND ENDGAME

Introduction

Chavan, U. M., and M. R. Borkar. "Observations on Cooperative Fishing, Use of Bait for Hunting, Propensity for Marigold Flowers and Sentient Behaviour in Mugger Crocodiles *Crocodylus palustris* (Lesson, 1831) of River Savitri at Mahad, Maharashtra, India." *Journal of Threatened Taxa* 15 (2023): 23750–62.

Dinosore

Chure, D. J., and M. A. Loewen. "Cranial Anatomy of *Allosaurus jimmadseni*, a New Species from the Lower Part of the Morrison Formation (Upper Jurassic) of Western North America." *PeerJ* 8 (2020): e7803.

Foth, C., S. W. Evers, B. Pabst, et al. "New Insights into the Lifestyle of *Allosaurus* (Dinosauria: Theropoda) Based on Another Specimen with Multiple Pathologies." *PeerJ* 3 (2015): e940.

Hanna, R. "Multiple Injury and Infection in a Sub-adult Theropod Dinosaur *Allosaurus fragilis* with Comparisons to Allosaur Pathology in the Cleveland-Lloyd Dinosaur Quarry Collection." *Journal of Vertebrate Palaeontology* 22, no. 1 (2022): 76–90.

Leiggi, P., and B. H. Breithaupt. *The Story of Big Al: Saving a Dinosaur for the Future*. Wyoming State Geological Survey Educational Series, 2009.

Marsh, A., and T. B. Rowe. "A Comprehensive Anatomical and Phylogenetic Evaluation of *Dilophosaurus wetherilli* (Dinosauria, Theropoda) with Descriptions of New Specimens from the Kayenta Formation of Northern Arizona." *Journal of Paleontology* 94 (2020): 1–103.

Senter, P., and S. L. Juengst. "Record-Breaking Pain: The Largest Number and Variety of Forelimb Bone Maladies in a Theropod Dinosaur." *PLOS One* 11 (2016): e0149140.

Senter, P., and C. R. Sullivan. "Forelimbs of the Theropod Dinosaur *Dilophosaurus wetherilli*: Range of motion, Influence of Paleopathology and Soft Tissues, and Description of a Distal Carpal Bone." *Palaeontologia Electronica* 22 (2019): 1–9.

Traumatic Amputation

Butler, R. J., A. M. Yates, O. W. M. Rauhut, and C. Foth. "A Pathological Tail in a Basal Sauropodomorph Dinosaur from South Africa: Evidence of traumatic amputation?" *Journal of Vertebrate Paleontology* 33 (2013): 224–28.

Yates, A. M. "A New Theropod Dinosaur from the Early Jurassic of South Africa and Its Implications for the Early Evolution of Theropods." *Palaeontologia Africana* 41 (2005): 105–22.

Deadly Decapitation

Klug, C., S. N. F. Spiekman, D. Bastiaans, B. Scheffold, and T. M. Scheyer. "The Marine Conservation Deposits of Monte San Giorgio (Switzerland, Italy): The Prototype of Triassic Black Shale Lagerstätten." *Swiss Journal of Palaeontology* 143 (2024): 11.

Spiekman, S. N. F., and E. Mujal. "Decapitation in the Long-Necked Triassic Marine Reptile *Tanystropheus*." *Current Biology* 33 (2023): R699–709.

Spiekman, S. N. F., J. M. Neenan, N. C. Fraser, et al. "Aquatic Habits and Niche Partitioning in the Extraordinarily Long-Necked Triassic Reptile *Tanystropheus*." *Current Biology* 30 (2020): 1–7.

Twisted Reptile

Burnham, D. A., B. M. Rothschild, J. P. Babiarz, and L. D. Martin. "Hemivertebrae as Pathology and as a Window to Behavior in the Fossil Record." *PalArch's Journal of Vertebrate Palaeontology* 10 (2013): 1–6.

Louis, M. L., J. M. Gennari, J. M. Vital, et al. "Congenital Scoliosis: A Frontal Plane Evaluation of 251 Operated Patients 14 Years Old or Older at Follow-Up." *Orthopaedics & Traumatology: Surgery & Research* 96 (2010): 741–47.

Sproule, D. M. "Scoliosis." *Encyclopedia of the Neurological Sciences* 4 (2014): 112–14.

Szczygielski, T., D. Dróżdż, D. Surmik, A. Kapuścińska, and B. M. Rothschild. "New Tomographic Contribution to Characterizing Mesosaurid Congenital Scoliosis." *PLOS One* 14 (2019): e0212416.

Szczygielski, T., D. Surmik, A. Kapuścinska, and B. M. Rothschild. "The Oldest Record of Aquatic Aamniote Congenital Scoliosis." *PLOS One* 12 (2017): e0185338.

Witzmann, F., P. Asbach, K. Remes, et al. "Vertebral Pathology in an Ornithopod Dinosaur: A Hemivertebra in *Dysalotosaurus lettowvorbecki* from the Jurassic of Tanzania." *The Anatomical Record: Advances in Integrative Anatomy and Evolutionary Biology* 291 (2008): 1149–55.

Witzmann, F., Y. Haridy, A. Hilger, I. Manke, and P. Asbach. "Rarity of Congenital Malformation and Deformity in the Fossil Record 474 of Vertebrates–A Non-Human Perspective." *International Journal of Paleopathology* 33 (2021): 30–42.

The Ice Ice Ice Baby

Fisher, D. C., A. N. Tikhonov, P. A. Kosintsev, et al. "Anatomy, Death, and Preservation of a Woolly Mammoth Calf, Yamal Peninsula, Northwest Siberia." *Quaternary International* 255 (2012): 94e105.

Fisher, D. C., E. A. Shirley, C. D. Whalen, et al. "X-ray Computed Tomography of Two Mammoth Calf Mummies." *Journal of Paleontology* 88 (2014): 664–75.

Kosintsev, P. A., E. G. Lapteva, S. S. Trofimova, et al. "The Intestinal Contents of a Baby Woolly Mammoth (*Mammuthus primigenius* Blumenbach, 1799) from the Yuribey River (Yamal Peninsula)." *Doklady Akademii Nauk* 432 (2010): 556–58.

Lopatin, A.V. "Yuka the Mammoth, a Frozen Mummy of a Young Female Woolly Mammoth from Oyogos." *Paleontological Journal* 55 (2021): 1270–74.

Lozhkin, A.V., and P. M. Anderson. "About the Age and Habitat of the Kirgilyakh Mammoth (Dima), Western Beringia." *Quaternary Science Reviews* 145 (2016): 104e116.

National Geographic. *Waking the Baby Mammoth*. Directed by Pierre Stone, 2009.

Papageorgopoulou, C., K. Link, and F. J. Rühli. "Histology of a Woolly Mammoth (*Mammuthus primigenius*) Preserved in Permafrost, Yamal Peninsula, Northwest Siberia." *Anatomical Record* 298 (2015): 1059–71.

Rountrey, A. N., D. C. Fisher, A. N. Tikhonov, et al. "Early Tooth Development, Gestation, and Season of Birth in Mammoths." *Quaternary International* 255 (2012): 196e205.

11. STRANGER THINGS

Introduction

Main, D. "A Far Cry from Normal: Amazonian Butterflies Drink Turtle Tears." NBC News, September 12, 2013. https://www.nbcnews.com/sciencemain/far-cry-normal-amazonian -butterflies-drink-turtle-tears-8c11138121.

Pryke, L. M. *Turtle*. Reaktion, 2020.

Bonehenge

Pryor, A. J. E., D. G. Beresford-Jones, A. E. Dudin, et al. "The Chronology and Function of a New Circular Mammoth-Bone Structure at Kostenki 11." *Antiquity* 94 (2020): 323–41.

Sablin, M., N. Reynolds, K. Iltsevich, and M. Germonpré. "The Epigravettian Site of Yudinovo, Russia: Mammoth Bone Structures as Ritualized Riddens." *Environmental Archaeology* 30, no. 1 (2023): 1–21.

Life's a Drag

Lomax, D. R., P. L. Falkingham, G. Schweigert, and A. P. Jiménez. "An 8.5 m Long Ammonite Drag Mark from the Upper Jurassic Solnhofen Lithographic Limestones, Germany." *PLOS One* 12 (2017): e0175426.

The Zombie Death Grip

Hughes D. P., T. Wappler, and C. C. Labandeira. "Ancient Death-Grip Leaf Scars Reveal Ant-Fungal Parasitism." *Biology Letters* 7 (2011): 67–70.

Knecht, R. J., A. Swain, J. S. Benner, et al. "Endophytic Ancestors of Modern Leaf Miners May Have Evolved in the Late Carboniferous." *New Phytologist* 240 (2023): 2050–57.

Ice Age Regeneration

Faerman, M., G. K. Bar-Gal, E. Boaretto, et al. "DNA Analysis of a 30,000-Year-Old *Urocitellus glacialis* from Northeastern Siberia Reveals Phylogenetic Relationships Between Ancient and Present-Day Arctic Ground Squirrels." *Scientific Reports* 7 (2017): 42639.

Gaglioti, B. V., B. M. Barnes, G. D. Zazula, A. B. Beaudoin, and M. J. Wooller. "Late Pleistocene Paleoecology of Arctic Ground Squirrel (*Urocitellus parryii*) Caches and Nests from Interior Alaska's Mammoth Steppe Ecosystem, USA." *Quaternary Research* 76 (2011): 373–82.

Kramina, T. E., E. P. Lakovenko, L. A. Koppel, et al. "Molecular Taxonomic Identification of a Silene Plant Regenerated from Late Pleistocene Fruit Material." *Wulfenia* 28 (2021): 29–50.

Oxelman, B, A. Petri, R. Elven, and G. Lazkov. "The Taxonomic Identity of the 30,000-y-old Plant Regenerated from Fruit Tissue Buried in Siberian Permafrost." *PNAS* 109 (2012): E2735.

Yashina, S., S. Gubin, S. Maksimovich, et al. "Regeneration of Whole Fertile Plants from 30,000-y-old Fruit Tissue Buried in Siberian Permafrost." *PNAS* 109 (2012): 4008–13.

Yashina, S., S. Gubin, S. Maksimovich, et al. "Reply to Oxelman et al.: On the Taxonomic Status of the Plants Regenerated from 30,000-y-old Fruit Tissue Buried in Siberian Permafrost." *PNAS* 109 (2012): E2736.

Zazula, G. D., D. G. Froese, S. A. Elias, S. Kuzmina, and R. W. Mathewes. "Arctic Ground Squirrels of the Mammoth-Steppe: Paleoecology of Late Pleistocene Middens (~24 000–29 450 14C yr BP), Yukon Territory, Canada." *Quaternary Science Reviews* 26 (2007): 979–1003.

Two Heads Better than One?

Buffetaut, E., J. Li, H. Tong, and H. Zhang. "A Two-Headed Reptile from the Cretaceous of China." *Biology Letters* 3 (2007): 80–81.

Index